Morphogenesis in Plants

Molecular Approaches

NATO ASI Series

Advanced Science Institutes Series

A series presenting the results of activities sponsored by the NATO Science Committee, which aims at the dissemination of advanced scientific and technological knowledge, with a view to strengthening links between scientific communities.

The series is published by an international board of publishers in conjunction with the NATO Scientific Affairs Division

A	**Life Sciences**	Plenum Publishing Corporation
B	**Physics**	New York and London
C	**Mathematical and Physical Sciences**	Kluwer Academic Publishers
D	**Behavioral and Social Sciences**	Dordrecht, Boston, and London
E	**Applied Sciences**	
F	**Computer and Systems Sciences**	Springer-Verlag
G	**Ecological Sciences**	Berlin, Heidelberg, New York, London,
H	**Cell Biology**	Paris, Tokyo, Hong Kong, and Barcelona
I	**Global Environmental Change**	

Recent Volumes in this Series

Volume 250 —Use of Biomarkers in Assessing Health and Environmental Impacts of Chemical Pollutants
edited by Curtis C. Travis

Volume 251 —Ion Flux in Pulmonary Vascular Control
edited by E. Kenneth Weir, Joseph R. Hume, and John T. Reeves

Volume 252 —Advances in Bacterial Paracrystalline Surface Layers
edited by Terry J. Beveridge and Susan F. Koval

Volume 253 —Morphogenesis in Plants: Molecular Approaches
edited by Kalliopi A. Roubelakis-Angelakis and Kiem Tran Thanh Van

Volume 254 —Slow Potential Changes in the Human Brain
edited by W. C. McCallum and S. H. Curry

Volume 255 —Quantitative Assessment in Epilepsy Care
edited by H. Meinardi, J. A. Cramer, G. Baker, and A. Martins da Silva

Volume 256 —Advances in the Biomechanics of the Hand and Wrist
edited by F. Schuind, K. N. An, W. P. Cooney III, and M. Garcia-Elias

Series A: Life Sciences

Morphogenesis in Plants

Molecular Approaches

Edited by

Kalliopi A. Roubelakis-Angelakis
University of Crete
Heraklion, Greece

and

Kiem Tran Thanh Van
CNRS
Gif-sur-Yvette, France

Springer Science+Business Media, LLC

Proceedings of a NATO Advanced Study Institute on
Plant Morphogenesis: Molecular Approaches,
held September 5–17, 1992,
in Crete, Greece

NATO-PCO-DATA BASE

The electronic index to the NATO ASI Series provides full bibliographical references (with keywords and/or abstracts) to more than 30,000 contributions from international scientists published in all sections of the NATO ASI Series. Access to the NATO-PCO-DATA BASE is possible in two ways:

—via online FILE 128 (NATO-PCO-DATA BASE) hosted by ESRIN, Via Galileo Galilei, I-00044 Frascati, Italy

—via CD-ROM "NATO Science and Technology Disk" with user-friendly retrieval software in English, French, and German (©WTV GmbH and DATAWARE Technologies, Inc. 1989). The CD-ROM also contains the AGARD Aerospace Database.

The CD-ROM can be ordered through any member of the Board of Publishers or through NATO-PCO, Overijse, Belgium.

Library of Congress Cataloging in Publication Data

Morphogenesis in plants: molecular approaches / edited by Kalliopi A. Roubelakis-Angelakis and Kiem Tran Thanh Van.

p. cm.—(NATO ASI series. A, Life sciences; vol. 253)

"Proceedings of a NATO Advanced Study Institute on Plant Morphogenesis: Molecular Approaches, held September 5–17, 1992, in Crete, Greece"—T.p. verso.

Includes bibliographical references and index.

ISBN 978-1-4899-1267-1

1. Plant morphogenesis—Congresses. 2. Plant molecular biology—Congresses. I. Roubelakis-Angelakis, Kalliopi A. (Kalliopi Apostolos) II. Tran Thanh Van, Kiem. III. NATO Advanced Study Institute on Plant Morphogenesis: Molecular Approaches (1992: Crete, Greece) IV. Series: NATO ASI series. Series A, Life sciences; v. 253.

QK665.M68 1993 93-20954
581.3′32—dc20 CIP

ISBN 978-1-4899-1267-1 ISBN 978-1-4899-1265-7 (eBook)
DOI 10.1007/978-1-4899-1265-7

PROLEGOMENA

Regeneration of plants from various types and sizes of explants has a history of almost a century. Also, plant regeneration from plant cells devoid of cell wall, the protoplasts, was firstly achieved in 1971. However, the success so far has been mostly based on trial and error efforts without studying much in depth the physiological, biochemical and molecular factor(s) and mechanism(s) which control at the molecular, cellular and explant level the morphogenic behaviour of protoplasts and plant explants.

The regeneration process includes two events which are of central importance for all living systems, including plants: cell division and morphogenesis. Not all plant species or even genotypes within each species show the same cell proliferating potential and morphogenic expression, at cell, callus and explant level, under identical culture conditions. Despite the painstaking long experimentation efforts, some of the most important plant genera for the mankind including most cereals, legumes and woody species, still show recalcitrance, that is the inability to express regenerating potential, either as cell division *in vitro* or as morphogenic response.

The application of molecular biology techniques in plants has significantly contributed in understanding physiological and biochemical functions at the gene level. The use of these techniques for elucidating the basic mechanism(s) of the processes involved in plant cell division and morphogenesis may eventually help to overcome the problem of recalcitrance. Consequently, the application of plant tissue culture and genetic engineering techniques may lead to the production of modified transgenic plants displaying new useful traits even in the case of recalcitrant species. Within this context, attention has been paid by the contributing authors on current status of knowledge concerning the physiological, biochemical and molecular events underlying plant morphogenesis processes.

The comprehensive reviews which are included in this volume were authored by the main lecturers of a NATO-Advanced Study Institute which was held in Heraklion, Crete from 5 to 17 of September 1992. The information in these reviews is directed at advanced students and professionals in the plant sciences: physiologists, biochemists, molecular biologists or those in horticultural, agricultural and forestry sciences. It is also intended that the book should serve as a text and guide to the literature for graduate level courses in Plant Morphogenesis or Plant Molecular Biology.

We would like to cordially thank the Scientific Affairs Division of NATO and especially Dr. L. da Cunha for having approved this ASI-Seminar and partially financing its implementation. Assistance of Ms. N. Clark and Mr. G. Safford, on behalf of the publishers, is gratefully acknowledged. We are also thankful to Ms. K. Efthimiopoulos for correcting the English. The production of the final copy was accomplished from author-submitted computer

Secreted Proteins in Plant Cell Cultures. 181
F.A. Van Engelen and S.C. De Vries

An Assessment of Possible Factors Contributing
to Recalcitrance of Plant Protoplasts. 201
K.A. Roubelakis-Angelakis

Gene Expression in Mesophyll Protoplasts. 221
Y. Meyer, Y Chartier, J. Grosset, I. Marty, C. Brugidou, P. Marinho, and R. Rivera

Antisense RNA in Plants: A Tool for Analysis and Suppression
of Gene Function. 237
M. Tabler

Phytoactive Polymers: New Synthetic Plant Growth Regulators. 259
A.M. Tsatsakis and M.I. Shtilman

Author Index 277

Subject Index 279

MORPHOGENIC SIGNALS AND MARKERS *IN VITRO* AND *IN VIVO*

Daphne J. Osborne

Oxford Research Unit
Open University
Foxcombe Hall
Boars Hill, Oxford OX1 5HR, UNITED KINGDOM

INTRODUCTION

To the best of our current knowledge, every living plant cell contains all the genetic information for programming a whole mature and reproductive adult individual. This implies that at no time during growth and development is this information lost or added to. Assuming this is so, then the basic question of morphogenesis concerns the nature and regulation of the expression of this information and the means by which it is directed to the development of an organized form and behaviour. Morphogenesis is thus the succession by which the genetic dictionary of the plant is revealed in an ordered sequence of differentiation and developmental events, all of which are constrained within a format that is the specific dictate of the species.

The assumption that genetic information is neither lost nor generated is of special importance to plants. For, in contrast to animals in which the special cells destined for reproduction are set aside early in development into specialized tissues distinct from somatic cells, in plants, the formation of reproductive cells represents a final expression of maturity in the cell lineages of what can be long-term somatic vegetative meristems.

The discussions in this review relate to the ways that plant cells retain total heritable information, but not withstanding, still develop ordered and potentially hierarchical states of complexity in tissue, organ and whole plant organization, so permitting the many and varied examples of morphogenic diversity. In particular, the pathways of information exchange and communication and the signals that may operate between cells, tissues and organs will be considered in relation to physical cues and morphogenesis directing molecules.

First, questions of organization in cells and tisssues will be addressed, to be followed by an appraisal of the relevance to whole plant systems. In reference to the role of hormones in genetic expression, emphasis will be laid upon a regulatory duet orchestrated by auxin and ethylene.

Morphogenesis in Plants, Edited by K.A. Roubelakis-Angelakis
and K. Tran Thanh Van, Plenum Press, New York, 1993

EARLY MORPHOGENESIS IN THE SINGLE CELL

Single cells dispersed freely in an aqueous environment are probably the closest approach to units independent of any influences from neighbouring cells of their own species. The zygotes of certain brown algae (*Fucus*, for example) are formed by the fusion of the male and female gametes after their liberation to the sea from the parent thallus. The free-floating egg cell thus is isolated from immediate signals from the parents. For the development of pattern and organised form there must however, be a generation of morphogenic fields to evoke new biochemical expressions in defined domains within the cell.

The prelude to such events is the establishment of a polarity within the cell. This occurs within several hours of fertilization and the visible expression of polarity is a protrusion that forms at the position of greatest wall weakness. The site designates the locus of future rhizoid growth and it is at this stage that the egg is said to have germinated. Since inputs from neighbouring cells are not involved, and the aqueous medium is chemically uniform, the important question is what signals are responsible for converting the apparently symmetrical egg cell to one of committed polarised growth.

Fucus eggs have been the subject of considerable study and it is clear that the critical forces acting upon the egg cell at this early stage are physical forces. Light, gravity and temperature all have a part to play. As Jaffe (1966) showed, the early positioning of the rhizoid protrusion can be initiated by the direction of light. Light generates ionic currents around the eggs which enter the cell at the relatively unlit side. The darkened side is then programmed as the site of rhizoid protrusion. The unilateral light is presumed detected by a blue photoreceptor though the means of signal transduction remains unknown (Brownlee, 1990).

By varying the Ca^{2+} composition of the artificial sea water into which the eggs are discharged and by the use of isotopic $^{45}Ca^{2+}$, Robinson and Jaffe (1975) showed that Ca^{2+} enters the unlit side and traverses the cell. With Ca^{2+} -sensitive microelectrodes Brownlee and Wood (1984) showed that the internal current thereby generates an internal Ca^{2+} flux and a gradient of 2 picoamps of Ca^{2+} current from one side of the cell to the other.

The position of high Ca^{2+} inward current also demarks a domain of cytosolic vesicle accumulation and the site of greatest wall weakness from where the rhizoid eventually protrudes and polarity of the germinating egg is established.

Blocking of the Ca^{2+} gradient blocks the development of polarity. Using cytochalasin D, Brawley and Robinson (1985) immobilised the actin fibres of the cytoskeleton with a resulting arresting of vesicle movement to the rhizoid pole. The inward Ca^{2+} current also was blocked and from this and other evidence Brawley and Robinson proposed that the entry of Ca^{2+} stimulated the localisation of F-actin at the presumptive rhizoid pole thereby enhancing vesicle migration to that region of the egg. Assuming the vesicles carry glucanhydrolases, amongst other protein components, then a domain of subsequent wall weakening and rhizoid emergence is delineated.

This is a model in which an initial physical signal (light), when transmitted for sufficient time (transient light signals can be reversed), will elicit electric currents and ionic gradients (in this case Ca^{2+}) which in turn are transduced to biochemically established morphogenic programmes.

The first event therefore is deemed to be the development of electrical polarity, with the second as a stabilization and entrainment of changes in cellular domains through the localised direction of biochemical pathways.

Not all the evidence supports the model, however. Kropf and Quatrano (1987) found no effect of removing Ca^{2+} from external solution on the development of polarity in *Fucus* or *Pelvetia* zygotes and in more recent experiments, Brownlee (1989) was unable to demonstrate a clear cytoplasmic gradient of Ca^{2+} within polarizing zygotes -though flooding them with injected calcium buffers successfully blocked polarity development (Speksnijder *et al.*, 1989).

Whilst the conclusion that a physical stimulus (light) is the initiator of electrical currents generated by Ca^{2+} as an intermediate in the transduction chain may well be correct, interactions between Ca^{2+} and other ion channels (K^+ or Na^+) during depolarization and repolarization of the plasma-membrane are more complex. Also the question of whether the rises in Ca^{2+} that are observed are through intracellular release or increased influx (or either/both) is still finally to be resolved (Brownlee and Taylor, 1992).

Clearly, the determination of polarity and the emergence of the rhizoid of the *Fucus* zygote is a single generated event, but in the more complex giant one-celled alga *Acetabularia mediterranea*, whorls of protrusions are generated at the growing tip every few days and these are reiterated at spaced intervals as the cell extends and expands. The final cell length may reach over 4 cm with a diameter of 400 µm. The whorls are generally shed and within months, growth of the apical dome is terminated in a reproductive structure, the cap. The spacing and timing are dependent both on temperature (Harrison *et al.*, 1981) and the Ca^{2+} concentration in the external medium. But of special interest is the evidence that whorl formation is preceded and determined by an annular pattern of Ca^{2+} distribution and of Ca^{2+} currents which has led to the speculation that morphogenesis is related to a Ca^{2+} determined pre-pattern (Harrison *et al.*, 1988).

A further important aspect of *Acetabularia* is that removal of the single nucleus situated at the rhizoid end does not preclude (if done after a precise stage of growth) the cap formation many weeks later (Haemmerling, 1963). All the information necessary for the production of the cap (cap-specific enzymes and cap-specific cell wall polysaccharides) must leave the nucleus long before the cap normally is formed and furthermore, this information remains stable within the cytoplasm for prolonged periods. The expression of morphogenic cues for the cap pre-pattern, (translation of the already transcribed genetic information) is therefore, controlled by mechanisms operating within the cytoplasm.

Hormones have been implicated in temporal regulation of cap formation (Vanden Driessche, 1984). Auxin (IAA 10^{-4} M) has been shown to enhance cap formation in whole cells once they have reached a certain stage of maturity and also in anucleate cells, provided that the mRNA's necessary for cap expression already have been transcribed and accumulated before removal of the nucleus. Such experiments serve to emphasise the developmental stage-dependent effect of the hormone and the part that it must play in post-transcriptional modulation of morphogenesis.

It is not difficult to conceive of ionic and molecular gradients between the poles of single plant cells with the attendant domains and morphogenic fields that these can engender and much has been acheived in the analysis of channel distribution at membrane surfaces by means of patch clamping. However, the cell wall must first be removed which precludes

mination by a variety of cytoplasmically generated field forces before a final permissive or repressive hormonal control mechanism becomes operative. However, as examples show, hormones also afford a feedback control to the electrical properties of the cell membrane, to the function of ion channels within the membrane and probably to the cytosolic levels of free Ca^{2+}. The interplay of such multistep controls provides a basis for cell specific or tissue specific gene expressions and a regulatory key for the differentiation of specific hormonally regulated target cells.

HORMONES AND ELECTRIC POTENTIALS

That hormones can interplay with ion currents and membrane potentials in cells from whole plant tissues and that these can be correlated with morphogenesis now seems without doubt. For example, on the isolation of protoplasts from the leaf mesophyll of an auxin-resistant mutant of tobacco (ie a mutant showing a much reduced rooting response to applied auxin compared with the wild genotype) Guern and his colleagues (Barbier-Brygoo *et al.*, 1990) showed that maximum hyperpolarization across the plasma membrane required an order of magnitude higher concentration of auxin (50 μM naphthalene acetic acid, NAA) than that of their wild type counterparts of 5 μM.

Conversely, protoplasts isolated from the mesophyll cells of leaves of tobacco plants transformed by the *rol B* gene of *Agrobacterium rhizogenes* (so that they became root initial over-producers) showed a 100-1000 fold increase in response with the transmembrane electrical potential difference at a maximum at 0.05-0.005 μM NAA. However, if the protoplasts were preincubated with the IgG for the auxin binding protein from corn coleoptiles and then tested for the electrical response, the differences between the protoplasts was reduced by at least an order of magnitude. It appears therefore, that the concentration dependencies of these electrical potential differences can be correlated with the density of auxin receptors upon the surface of the plasmamembranes of the different types of protoplasts. But, it also shows that the root forming capacity of these different tobacco genotypes can be linked equally well to the concentration dependence of the electrical potential differences that an auxin can evoke.

HORMONES AND ION CHANNELS

Until recently the direct effect of a hormone upon an ion channel in a highly differentiated plant cell had not been clearly demonstrated. Last year however, Marten *et al.* (1991) reported such control in peels from the epidermis of *Vicia faba* leaves in which stomatal opening could be elicited either by light or by auxins (IAA, NAA or 2,4-D at ca. 10^{-5} M). Using patch clamp techniques with guard cell protoplasts from these peels, they showed the usual auxin-induced increase in proton secretion (activation of H^+ -ATPases) 3-10 min after the auxin application, but also, an earlier rapid and reversible (9-10ms at -40 to -60 mV) activation of inward and outward Cl^- currents. An inhibition of Cl^- (in) current before a peak in the Cl^- (out) current and the accompanying activation potential shifts to more hyperpolarized potentials. No other hormones induced this response. Comparisons of whole cell currents and single channel activity in a membrane patch excised from the same cell showed that macroscopic currents (activation shift of ca.-20mV) are generated by the auxin-sensitive anion channels. In similar experiments, K^+ currents were not affected by auxins, indicating that the K^+ (in) channels of guard cells at least, may lack these specific cation target sites for auxin. One may speculate therefore, whether the gating of specific ion channels by specific hormonal substances could be a general event and whether it is part of the complex orchestration of morphogenic control.

STABILITY OF MORPHOGENIC STATES

Let us consider further the cell cluster that has acquired the distinctive polar characteristics of a somatic embryo. Racusen and Schiavone (1988) showed that as many as 5% of the total proteins extracted from the root pole or the shoot pole of carrot embryoids up to 1mm in length and fractionated by 2-dimensional SDS-PAGE, can be recognised as unique to that part of the embryo. Polarity of protein formation and distribution therefore, has already taken place. A hormone involvement in polarized growth is also evident at this stage, for as Schiavione (1988) found if the apical end of the embryo is removed, the root pole undergoes rapid growth whilst restoring the apex, (or instead, substituting a donor block of auxin) returns root growth to its normal rate.

By using miniaturized surgical techniques, Schiavione and Racusen (1990) then showed that torpedo stage embryos of carrot, transected so that from 25 to 90% of the apical pole was present, would regenerate a new root pole at the cut end. In contrast, excised root poles containing more than 10% of the length of the embryoid would continue only as roots. Using 2-dimensional electrophoretic fractionation of ^{35}S-methionine-labeled proteins extracted from single embryos, Racusen and Schavione (1990) found 3 of the root pole proteins on the electrophoretograms of shoot pole sections that underwent root-regeneration. Since reassembly of severed pieces within 24h of cutting permitted the normal pathway of embryogenesis to be resumed (Schavione, 1988) with both root and shoot proteins remaining restricted to their respective poles, the appearance of the root pole proteins in the root regenerating shoot poles is unlikely to be the result of wounding. Rather, it can be considered as the temporal and spatial expression of cell or tissue specific genes correlating with well coordinated pattern forming routines. Such results demonstrate that in carrot cultures at least, the cell from a shoot pole enjoys a longer period of flexibility of committment than a cell from the root pole. This suggests that the establishment of the shoot cell pathway is determined relatively late in morphogenesis compared with the root.

The meristems of somatic embryos bear a close resemblance to those of embryos formed *in ovulo* - both grow to produce morphologically and biochemically similar meristems on the growing plant. An early switch that constrains future differentiation in the whole plant meristem is the cell lineage distinction that delimits cells of the epidermis (the L1 layer) from those of the inner parenchyma layers (L2 and L3) as described by Poethig (1984) (Fig. 1). This has special importance in the consideration of abscission zone chimaeras discussed later.

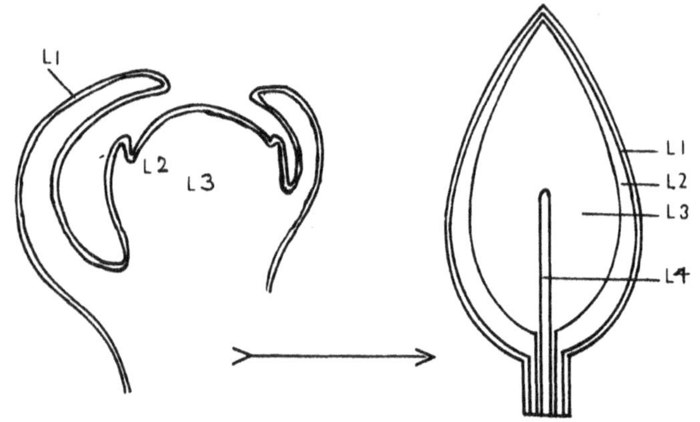

Figure 1. Cell lineages in apical meristems and developing leaves.

Within these constraints organ formation from a committed meristem on the whole plant is subject to temporal control in an apparently similar way to that of the somatic embryo. In *Pisum*, for example, Sachs (1969) showed that when the youngest visible leaf primordium (less than 70 µm long) was cut in half and one of the halves removed, a complete new entire leaf was regenerated. If he bisected the primordium, then two new and entire leaves were formed. But when older leaf primordia (greater than 70 µm long) were so treated, complete regeneration failed to occur depending upon which part had been excised. This indicates that as a primordium becomes more differentiated, the potential pathways for further differentiation are restricted.

What then is the biochemical nature of these committed states and the mechanisms by which prepatterned information can be directed along precise developmental pathways, some of which lead to stages of differentiation that are apparently terminal.

THE CONCEPT OF THE TARGET CELL

The first part of this review addressed ways in which cells or tissues might establish degrees of determination as a prelude to the further differentiation of the plant body and eventually to states of differentiation which are terminal. The expression of flowering in annual plants is an example of such terminal differentiation. Transmissable physiological signals in the vegetative apex initiate the floral conversion. As with the specific marker proteins of the root and shoot poles demonstrated by Racusen and Schavione (1988), floral abundant polypeptides and their cDNA's have been isolated from all the floral parts of tomato mutant *anantha* (a flower-bud arrested mutant that proliferates flower primordia) and in all the floral parts of the wild genotype (Shahar *et al.*, 1992). Interestingly, a 66.3 kDa polyphenoloxidase (P2) which is one of the molecular markers for the floral tomato tissue, is also expressed in the vegetative phase but only to any extent by the palisade cells and epidermal trichomes. This selective and cell-specific distribution of gene products illustrates the strict control exerted on the endogenous gene pool by cells of particular determined and committed states. It implies also that cells of particular determined and committed states will perceive stimuli, transmit signals and be directed along differentiation pathways as dictated by their state of competence to perceive and respond to those stimuli. The stimuli may be external (light, gravity, wounding etc.) or internal (hormonal, morphogenic regulators, intertissue mechanical pressures). The target cell concept portrays how each living cell in the plant body will be a target cell for one or other of these stimuli and respond in its own pre-programmed way (Osborne, 1984). One of the challenges for morphogenesis studies at the whole plant level is the unraveling of the molecular basis for the competence of any target cell type to respond to a given stimulus, and further, to identify the specific proteins or molecular determinants expressed that constitute the operational markers of the specific state of competence.

EVIDENCE FOR TARGET CELL TYPES

The cells of aleurone tissue in graminaceous seed are classic examples of target cells. Gibberellin, produced by germinating embryo axes, passes to the surrounding cells of the aleurone and there induces transcription (and translation) of mRNAs from genes that code for α-amylase. The events in the aleurone are held repressed by an appropriate concentration of another hormone, abscisic acid (Jacobsen and Beach, 1985). No other cells of the seed synthesise this protein. Whilst much is understood of the molecular biology of α-amylase induction far less is understood of the controls that determine the differentiation of aleurone cells from their endosperm or axis neighbours or of the nature of their molecular markers.

The cells of abscission zones, however, have afforded a fertile field for such developmental studies (Osborne, 1989, 1991), for not only is a zone-associated isozyme of an endo- β-1,4-glucanhydrolase (a cellulase, pI 9.5) induced by ethylene and repressed by auxin, but diagnostic marker peptides have been identified for zone cells of some species. Also, clues to the control of abscission cell differentiation by hormonal gradients have provided new insights into cell determination.

THE ABSCISSION ZONE

Molecular Markers

Cells of two of the most studied abscission zones of higher plants, those of the distal pulvinus of the primary leaf of the dwarf bean (*Phaseolus vulgaris*) and the compound leaf of the elderberry (*Sambucus nigra*), both have been shown to be distinct biochemically from their neighbour cells on either side. Each possesses peptides that either are not detectible in the non-zone cells (as in *Sambucus*, McManus and Osborne, 1990a) or are expressed quantitatively as an enriched proportion of the total protein complement (as in the zones of *P. vulgaris*, McManus and Osborne, 1990b). By SDS-PAGE fractionation of extracted proteins

It is tempting to speculate that the burst of ethylene associated with the growth of lateral root initials and the level of auxin produced in the young primary leaves together provide signals for the positional differentiation of the abscission zone target cells at the base of each pulvinus. Once differentiated the zone cells remain in a competent target state throughout the life-time of the leaf and are induced to separate *in vivo* only when the auxin gradient from the leaf blade falls below a critical level and the ethylene production of the adjoining senescent pulvinus rises.

In the bean, separating abscission zone cells liberate phloroglucinol-staining pentose sugars from the walls. It is significant therefore, that only those axes isolated at 120h or later from the germinating seed exhibit this phloroglucinol-positive staining in the wall at sites of cell from cell separation. This may mean that the differentiation of zone target cells involves the expression of new genes that code for enzymes (or proteins) that act (or must be present) in both the cytoplasm and the cell wall.

Abscission-Lacking Mutants

Chimaeras of the abscission lacking mutant line of tomato (jointless) with the wild type *Lycopersicon peruvianum* studied by Szymkowiak and Sussex (1989) have shown that the tissue that comprises L2 and L3 layers dictates the competence for abscission. If *jointless* cells constitute the inner tissue then competent zone target cells are not produced. In the mutant of *Lupinus angustifolium* (*Abs*$^-$) in which leaf and pod abscission does not occur, there is only a minimal presence of an active β-1,4-glucanhydrolase in the tissues of the leaflet bases following exposure to ethylene. Ethylene will however, induce increases (to the same extent) in both mutant and wild type genetic lines of a cytoplasmic and a wall-associated β-glucosidase activity in the petiole, and in a wall-associated β-glucosidase at the junction of the leaflet bases (the abscission zone region) (Osborne and Thompson, 1992). Clearly, the *Abs*$^-$ plants are not unresponsive to ethylene. So the failure to undergo abscission, even with high levels (1000 $\mu l.l^{-1}$) of ethylene may be due to a mutation in the cellulase gene or in its promoter.

Although we do not know the *in vivo* substrate for the endo- β-1,4-glucanhydrolase, the close association of the induction of its activity at the site of zone cell separation and (in the bean) the phloroglucinol staining of the separating walls make it difficult not to consider the enzyme as an essential player in the wall loosening and cell separation process. There are now many obvious approaches for exploring the lupin mutant and the β-1,4-glucanhydrolase (cellulase) gene which we currently are developing. But the following experiments with bean suggest alternate possibilities to those of a modified cellulase gene. The perception and response of the zone target cells in *Abs*$^-$, (and in other abscission-less mutants) may be normal but they may fail to receive essential inductive signals from the associated tissues of the abscission zone.

Intertissue Signalling in Abscission

As Reid *et al.* (1990) have shown already for the distal leaf abscission zone of bean, the pI 9.5 glucanhydrolase activity arises first in the vascular tissue but appears thereafter throughout the abscission region, progressing outwards from the stele. We

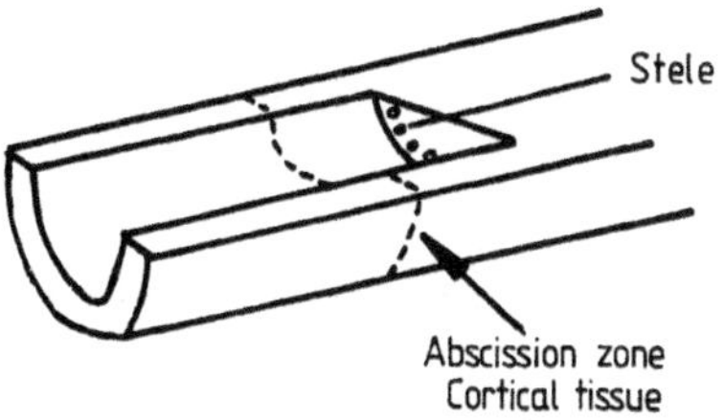

Figure 3. Inter-tissue signalling. Diagram of abscission region of bean explant with the stelar tissue removed.

have found that surgical removal, or dissociation of the stele from the cortex prevents both the glucanhydrolase development in the cortex as well as cell separation at the zone, even in the presence of ethylene (Osborne and Thompson, 1992). The expression of zone cell competence is therefore not attributable to the recognition of ethylene alone but requires also a signal that ethylene elicits in the vascular-associated tissues (Fig. 3).

However, if the stelar tissue is removed by microsurgery 24h after exposure to ethylene, but before the glucanhydrolase is present in the cortical tissue, a cortical glucanhydrolase will still be produced in the zone cortex after a further 48h in ethylene and the target cortical cells of the zone will separate normally at 96h. An ethylene-induced signal therefore passes from the stelar tissue to the zone cells. This signal is permissive to the endogenous production of the cortical glucanhydrolase. The signal cannot be substituted for by ethylene and to date only direct contact between stelar tissue and cortical tissue (by transplanting another stele into the cortex) has provided the effective route for the transmission of the signal. The transmitted signal also must become localized strictly within the cortex, for only cells in the immediate area of contact with the substituted stele are induced to separate or to produce the glucanhydrolase.

CONCLUSIONS FOR ABSCISSION

The positional differentiation of abscission zone target cells with a competence to separate in response to the initiating signal ethylene appears to be a relatively late event in leaf development. In the axes of germinating seedlings of bean, differentiation of zone cells in the immature primary leaf follows a period of high ethylene production by the root. This occurs at a precise stage in root morphogenesis when lateral roots are initiated, suggesting a developmental control by gradients of two sets of hormones, one from the shoot (auxin?) and one from the root (ethylene) acting in concert. Secondary abscission zones can be induced in explants of the petioles of mature leaves of the bean. The cells of the cortical parenchyma derived from L2 and L3 layers, can undergo transdifferentiation into competent zone cells and the advent and position of the conversion is also subject to regulation by the levels of both auxin and ethylene to which the tissue is exposed (Fig. 4).

Production of a β-1,4-glucanhydrolase enzyme is linked closely to the ability of the cells to separate and in the mutant *Abs*$^-$ lupin, in which this gene is not expressed as an active enzyme, no abscission occurs. Although ethylene will initiate the abscission process,

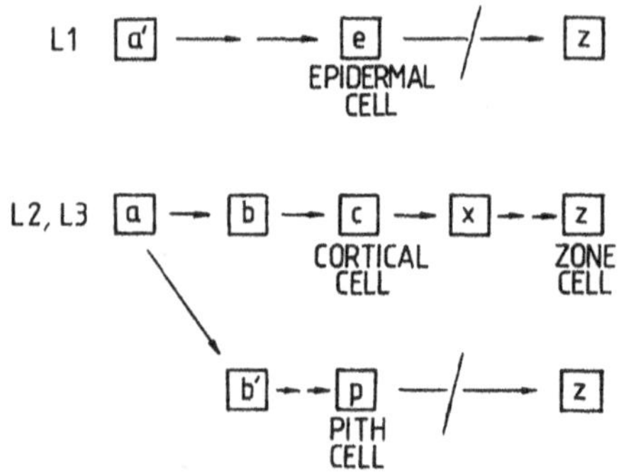

Figure 4. Cell lineages and the capacity for transdifferentiation to determined zone cells status.

target cells with a competence to separate in response to ethylene must be differentiated for abscission to occur. Inter-tissue signaling between cells of the stele and the cortex appears also to be a necessary component for cortical zone cells to achieve both the production of the glucanhydrolase and cell separation. A failure to complete abscission in response to ethylene should result if only one of the intercommunicating cell types fails to achieve the appropriate target state of gene expression. Work proceds to resolve these questions.

MORPHOGENESIS IN THE FUTURE

As always, there are more questions than answers and the more answers we have, the more questions we ask. Certain rules in morphogenesis do however, seem clear. A plant cell can sense and respond with remarkable rapidity (in ms) to environmental and hormonal stimuli by changes in ionic fluxes, electrical currents and transmembrane potentials. In response to ethylene for example, microtubule arrays of the cytoskeleton in pea stem cells can be reoriented through 90° within 30 min, thereby changing the internal polarity of the cell (Roberts *et al.*, 1985). Sensitive molecular techniques can detect auxin-induced changes in specific mRNA transcriptions in maize coleoptile cells within 10 min (Zurfluh and Guilfoyle, 1982). Elegant microsurgery tells us that stimuli must be sustained and signals perceived for minimal periods of time to reach thresholds of response before a developmental change can be entrained and a new state of target cell competence attained. It seems that resolving the interplay and timing of the critical molecular and biochemical events that direct morphogenesis will continue to challenge the developmental biologist for a long time to come.

ACKNOWLEDGEMENTS

I am indebted to Mrs Janice Henderson for preparing all the diagrams and Miss Beverley Wood for processing the manuscript.

REFERENCES

Barbier-Brygoo H, Maurel C, Shen WH, Ephritikhine G, Delbarre A, Guern J (1990) Use of mutants and transformed plants to study the action of auxins. *In:* Hormone Perception and Signal Transduction in Animals and Plants, Roberts J, Kirk C, Venis M (eds), *Soc Exp Biol Symp* XLIV, Cambridge UK, pp 67-77.

Brawley SH, Robinson KR (1985) Cytochalasin treatment disrupts the endogenous currents associated with cell polarization in fucoid zygotes: studies of the role of F-actin in embryogenesis. *J Cell Biol* 100: 1173-1184.

Brownlee C, Wood WJ (1984) A gradient of cytoplasmic free calcium in growing rhizoid cells of *Fucus serratus*. *Nature* 320: 624-626.

Brownlee C (1989) Visualizing cytoplasmic calcium in polarizing zygotes and growing rhizoids of *Fucus serratus*. *Biol Bull* 176: 14-17.

Brownlee C (1990) Light and development: cellular and molecular aspects of photomorphogenesis in brown algae. *In*: Light and Life in the Sea, Herring PJ, Campbell AR, Whitfield M, Maddock L (eds), Cambridge University Press, Cambridge, UK, pp 115-126.

Brownlee C, Taylor AR (1992) Cytoplasmic calcium, calcium currents and fluxes during fertilization and early development of *Fucus serratus* Progress. *In*: Plant Growth Regulation, Karssen CM, van Loon LC, Vreugdenhil D (eds), Kluwer Academic Publishers, Netherlands, pp 651-662.

Goodwin BC (1985) What are the causes of morphogenesis? *Bio Essays* 3: 32-36.

Goodwin BC (1990) Structurals in biology. *Sci Progress Oxford* 74: 227-244.

Gorst J, Overall, Wernicke RL and W (1987) Ionic currents traversing cell clusters from carrot suspension cultures reveal perpetuation of morphogenetic potential as distinct from induction of embryogenesis. *Cell Differentiation* 21: 101-109.

Haemmerling J (1963) Nucleo-cytoplasmic interactions in *Acetabularia* and other cells. *Annu Rev Plant Physiol* 14: 65-92.

Harrison LG, Snell J, Verdi R, Vogt DE, Zeiss GD, Green BR (1981) Hair morphogenesis in *Acetabularia mediterranea* : temperature-dependent spacing and models of morphogen waves. *Protoplasma* 106: 211-221.

Harrison LG, Graham KT, Lakowski BC (1988) Calcium localization during *Acetabularia* whorl formation: evidence supporting a two-stage hierarchical mechanism. *Development* 104: 255-262.

Jacobsen JV, Beach LR (1985) Control of transcription of α-amylase and rRNA genes in aleurone protoplasts by gibberellin and abscisic acid. *Nature* 316: 275-277.

Jaffe LF (1966) Electrical currents through the developing *Fucus* egg. *Proc Natl Acad Sci USA* 56: 1102-1109.

Kropf DL, Quatrano RS (1987) Localization of membrane-associated calcium during development of fucoid algae using chlorotetracycline. *Planta* 171: 158- 170.

Marten I, Lohse G, Hedrich R (1991) Plant growth hormones control voltage- dependent activity of anion channels in plasmamembrane of guard cells. *Nature* 353: 758-762.

McManus MT, Osborne DJ (1990a) Identification of polypeptides specific to rachis abscission zone cells of *Sambucus nigra*. *Physiol Plant* 79: 471-478.

McManus MT, Osborne DJ (1990b) Evidence for the preferential expression of particular polypeptides in leaf abscission zones of the bean *Phaseolus vulgaris* L.J. *Plant Physiol* 136: 391-397.

McManus MT, McKeating J, Secher DS, Osborne DJ, Ashford D, Dwek RA, Rademacher TW (1988)

Identification of a monoclonal antibody to abscission tissue that recognises xylose/fucose-containing N-linked oligosaccharides from higher plants. *Planta* 175: 506-512.

McManus MT, Osborne DJ (1991) Identification and characterization of an ionically-bound cell wall glycoprotein expressed preferentially in the leaf rachis abscission zone of *Sambucus nigra* l. *J Plant Physiol* 138: 63-67.

Osborne DJ (1984) Concepts of target cells in plant differentiation. *Cell Differentiation* 14: 161-169.

Osborne DJ (1989) Abscission. *Crit Rev Plant Sci* 8: 103-129.

Osborne DJ (1991) Ethylene in leaf ontogeny and abscission. *In*: The Plant Hormone Ethylene, Mattoo AK, Suttle JC (eds), CRC Press, Boca Raton, Florida, USA, pp 193-214.

Osborne DJ, McManus MT, Webb J (1985) Target cells for ethylene action. *In*: Ethylene and Plant Development, Roberts JA, Tucker GA (eds), Butterworths, London, UK, pp 197-212.

Osborne DJ, McManus MT (1986) Flexibility and commitment in plant cells during development. *Current Topics in Dev Biol* 20: 383-396.

Osborne DJ, Thompson DS (1992) Target or non-target: hormonal signal perception and response in the determination of cell performance. *In*: Progress in Plant Growth Regulation, Karssen CM, van Loon LC, Vreugdenhil D (eds), pp 237-247.

Poethig S (1984) Cellular parameters of leaf morphogenesis in maize and tobacco. *In*: Contemporary Problems in Plant Anatomy, White RA, Dickinson WC (eds), Academic Press, New York, USA, pp 235-259.

Racusen RH, Schiavone FM (1988) Detection of spatially- and stage-specific proteins in extracts from single embryos of the domesticated carrot. *Development* 103: 665-674.

Racusen RH, Schiavone F (1990) Positional cues and differential gene expression in somatic embryos of higher plants. *Cell Differentiation and Development* 30: 159-169.

Reid PD, del Campillo E, Lewis N (1990) Anatomical changes and immunolocalization of cellulase during abscission as observed on nitrocellulose tissue prints. *Plant Physiol* 93: 160-165.

Roberts IN, Lloyd, CW, Roberts K (1985) Ethylene-induced microtubule reorientations: mediation by helical arrays. *Planta* 164: 439-447.

Robinson KR, Jaffe LF (1975) Polarizing fucoid eggs drive a calcium current through themselves. *Science* 187: 70-72.

Sachs T (1969) Regeneration experiments on the determination of the form of leaves. *Israel J Bot* 18: 21-30.

Schiavone FM (1988) Microamputation of somatic embryos of the domestic carrot reveals apical control of axis elongation and root regeneration. *Development* 103: 657-664.

Schiavone FM, Racusen RH (1990) Microsurgery reveals regional capabilities for pattern reestablishment in somatic carrot embryos. *Dev Biol* 141: 211-219.

Shacklock PS, Read ND, Trewavas AJ (1992) Cytosolic free calcium mediates red light-induced photomorphogenesis. *Nature* 358: 753-755.

Shahar T, Hennig N, Gutfinger T, Hareven D, Lifschitz E (1992) The tomato 66.3-kD polyphenoloxidase gene: molecular identification and developmental expression. *The Plant Cell* 135-147.

Speksnijder JA, Miller AL, Weisenseel MH, Chen T-H, Jaffe LF (1989) Calcium buffer injections block fucoid egg development by facilitating calcium diffusion. *Proc Natl Acad Sci* 86: 6607-6611.

Szymkowiak EJ, Sussex IM (1989) Chimeric analysis of cell layer interactions during development of the flower pedicel abscission zone. *In*: Cell Separation in Plants, Osborne DJ, Jackson MB (eds), Springer-Verlag, Heidelberg, Germany, pp 363-368.

IN VITRO ORGANOGENESIS AND SOMATIC EMBRYOGENESIS: PHYSIOLOGICAL AND BIOCHEMICAL ASPECTS

Trevor A. Thorpe

Plant Physiology Research Group
Department of Biological Sciences
University of Calgary
Calgary, Alberta T2N 1N4, CANADA

INTRODUCTION

The study of plant morphogenesis is one area of research with which tissue culture long has been associated, and one in which the use of the *in vitro* technology has made significant contributions to both fundamental knowledge and application (Thorpe, 1990). In particular, our understanding of the processes of xylogenesis (cytodifferentiation), somatic organogenesis and somatic embryogenesis has been clearly enhanced through use of tissue cultures. These processes enable cells and tissues, which are quiescent or committed to some function or pathway of development to be channelled into some different aspect of organized *de novo* development.

This paper is concerned with two aspects of organized development, namely *de novo* organogenesis and somatic embryogenesis. Organogenesis is the process by which cells and tissues are forced to undergo changes which lead to the production of a unipolar structure, namely, a shoot or root primordium, whose vascular system is often connected to the parent tissues. In contrast, somatic embryogenesis leads to the production of a bipolar structure containing a root/shoot axis, with a closed independent vascular system. Both can occur directly on the explant or indirectly via callus formation. Callus is wound parenchyma tissue, which can be maintained more or less indefinitely in solid and/or liquid conditions, when it is subcultured at regular intervals onto new medium under aseptic conditions. Callus has proven to be a very useful material for the study of organized development, although several problems exist with respect to its use (Thorpe, 1982). In addition to callus, virtually any plant part can serve as inoculum, particularly for organogenesis. Such explants include stem and root segments, leaf sections, inflorescence sections, seedling parts such as cotyledons, hypocotyls, etc. and immature and mature seed embryos. Thin cell layers are also a powerful experimental system to study organogenesis (Tran Thanh Van, 1980; and this volume), but will not be discussed in this review.

Morphogenesis in Plants, Edited by K.A. Roubelakis-Angelakis
and K. Tran Thanh Van, Plenum Press, New York, 1993

The earliest reports on controlled organogenesis *in vitro* were by White (1939), who obtained shoots on callus of a tobacco hybrid, and by Nobecourt (1939), who observed root formation in carrot callus. The finding of White was confirmed and extended by Skoog (1944), who showed that auxin could stimulate rooting and inhibit shoot formation, but that the inhibitory effects of auxin could be reduced by increasing the concentration of other medium components, such as inorganic phosphate and adenine. Further studies by Skoog and his associates culminated in the now classical conclusions that organ formation resulted from quantitative interactions between growth regulators, particularly auxin and cytokinin, and other metabolites (Skoog and Miller, 1957). Studies with tobacco callus showed that a high ratio of auxin to cytokinin in the medium favored root formation, the reverse shoot formation, and that intermediate ratios promoted callus proliferation. Although this approach cannot be demonstrated universally, the concept played a pivotal role in shifting research away from the search for specific morphogens for different types of morphogenesis (although the quest for florigen has not yet been abandoned completely) and placed emphasis on the chemical interaction of substances in the regulation of growth and development (Brown and Thorpe, 1986). As a result of this approach several thousand plant species have been reported to form shoots and/or roots *in vitro*. Regeneration of plantlets via organogenesis, which is a multistaged process (Thorpe and Patel, 1984), is the widest application of tissue culture technology.

In contrast to the above, the earliest reports on controlled somatic embryogenesis *in vitro* were in 1958 with carrot (Reinert, 1958; Steward *et al.*, 1958). The process occurs naturally in a wide range of species from both reproductive and somatic tissues (Tisserat *et al.*, 1979), but can be induced in fewer species *in vitro*. Somatic embryogenesis can be induced *in vitro* from callus, cell suspension and protoplast cultures or directly from cells of organized structures such as a stem segment or zygotic embryo. As an approach to plantlet regeneration, somatic embryogenesis has several advantages over organogenesis, although not as readily achieved (Thorpe, 1988a). The advantages include the efficiency of the process (the formation of plantlets in fewer steps, with a concomitant reduction in labor, time and cost), the potential for the production of much higher number of plantlets, and the morphological and cytological uniformity of the plantlets. Today about 150 species from both angiosperms and gymnosperms have been reported to undergo somatic embryogenesis.

In this review, as indicated by the title, major emphasis will be placed on the physiological and biochemical aspects of organogenesis and somatic embryogenesis. This will be achieved by the review of work carried out on organogenesis in tobacco callus and excised cotyledons of radiata pine mainly by the author and his associates, and on somatic embryogenesis in carrot (the work of several labs) and white spruce. Besides representing herbaceous angiosperm and gymnosperm species, the material allows for morphogenesis indirectly from callus or directly from the explant. Under these circumstances, it should be possible to determine those events which are causally related to organized development per se. Since morphogenesis *in vitro* is dependent on the proper selection of the explant, the medium and the culture physical environment and since there is a strong interaction between these components the tissue culture systems used and the manipulation of the processes will be briefly discussed. As well, the manipulations carried out above are ultimately reflected in structural/developmental changes in the cells and tissues leading to appearance of the unipolar shoot or root primordium or the bipolar somatic embryo. These latter structures arise from preceding physiological, biochemical and molecular events occurring in the tissue. Thus, to fully appreciate the relevance of these events for organized development, the structural and developmental sequence must be determined.

TISSUE CULTURE SYSTEMS AND THEIR MANIPULATION

Tobacco (*Nicotiana tabacum* L. cv. W38) callus is generally initiated from stem pith-cambium sections and maintained on MS salts (Murashige and Skoog, 1962), supplemented with White's organics (White, 1943), inositol (100 mg l^{-1}), IAA (10 μM), kinetin (2 μM), Difco casamina acids (1 g l^{-1}) in the dark. For experimental purposes, 60-80 mg pieces taken from 4-7 week-old callus, were placed on shoot-forming (SF), non-shoot- forming (NSF) or shoot-forming plus gibberellin (SFG) media (Thorpe and Murashige, 1970). The SF medium consisted of MS salts supplemented with thiamine-HCl (1.2 μM), myo-inositol (100 mg l^{-1}), IAA (10 μM), kinetin (10 μM), adenine sulfate (0.98 mM), L-tyrosine (0.55 mM), $NaH_2PO_4.H_2O$ (1.2 mM), and sucrose (3% w/v). NSF medium consisted of MS salts with the above levels of thiamine-HCl, myo-inositol, IAA and sucrose, and with kinetin (2 μM). The SFG medium was the same as the SF medium, but contained in addition gibberellic acid (GA_3, 50 μM), which represses organogenesis (Murashige, 1961). All supplements were usually added to the medium prior to autoclaving, including GA_3 (Thorpe and Meier, 1973); and all media were solidified with Difco Bacto-agar® (0.9%, w/v). Most experiments were carried out in the dark at ca 27°C; when tissues were grown in the light, a 16 h photoperiod at a photon fluence rate of ca 80 $\mu mol.m^{-2}.s^{-1}$ PAR from wide spectrum fluorescent lamps was used. Generally 3 to 4 pieces of callus were placed in each 125 ml Erlenmeyer flask containing 50 ml medium. Flasks usually were closed with foam bungs.

The radiata pine (*Pinus radiata* D. Don) explant system consists of cotyledons excised from aseptic, dark-germinated seed approximately one day after radicle emergence (Aitken *et al.*, 1981). At this stage, the radicle is between 0.5 and 2.0 cm long and the cotyledons between 3 and 5 mm. Younger or older cotyledons lead to a reduced production of adventitious shoots (Aitken *et al.*, 1981; Aitken-Christie *et al.*, 1985). Each embryo had seven to ten cotyledons, and about 90% of the seeds behaved similarly in culture. The excised cotyledons are generally cultured in sterile plastic petri dishes on a modified Schenk and Hildebrandt medium (Reilly and Washer, 1977) containing N^6-benzyladenine (BA, 25 μM) and sucrose (3%, w/v) in the light (16 h at ca 80 $mol.m^{-2}.s^{-1}$ from wide spectrum fluorescent tubes) at 27±1°C. Cotyledons cultured in the absence of BA and old cotyledons (cotyledons excised 5 days post-germination) plus BA served as NSF controls.

Carrot (*Daucus carota* L) callus can be initiated successfully from any part of the carrot, although petiole and taproot segments are the most commonly used explants (Ammirato, 1984). These are usually placed on MS basal salts and vitamins containing 1-4.5 μM 2,4-dichlorophenoxyacetic acid (2,4-D). To initiate embryogenic suspension cultures, callus is transferred to liquid medium of the same composition using Erlenmeyer flasks or a Steward auxophyton (Steward and Shantz, 1956). Suspensions with various sizes of cell clusters are subcultured every 14-21 days. To obtain fine embryogenic cell suspensions and some degree of uniformity for experimentation, sieving is carried out using stainless steel or nylon mesh, or glass beads (Ammirato, 1984). Komamine and his associates established a high frequency and synchronous somatic embryogenic system in carrot by selecting competent single cells. These cells were small, round and cytoplasmic- rich, and were obtained by sieving with a nylon screen, density gradient centrifugation and manual selection (Nomura and Komamine, 1985). Selected cells (state 0) were pretreated with auxin (2,4-D, 0.05 μM), zeatin (1 μM), mannitol (3 mM) and high O_2 (40%), and then transferred to auxin-free medium in which the early stages of somatic embryo development occurred synchronously at about 85-90% frequency.

White spruce [*Picea glauca* (Moench) Voss] embryogenic tissue was obtained from immature zygotic embryos (1.5-2 mm) cultured on von Arnold and Eriksson s (AE, 1981) salts, vitamins and carbohydrates, supplemented with casein hydrolysate (0.5 g l^{-1}), glutamine (0.1 g l^{-1}), myo-inositol (0.1 g l^{-1}), along with picloram (4-amino-3,5,6- trichloropicolinic acid, 5 or 10 μM) and BA (5 μM) (Lu and Thorpe, 1987). Embryogenic tissue was subcultured on AE medium or transferred to DCR medium (Gupta and Durzan, 1985) both containing 2,4-D (10 μM) and BA (2 μM) and subcultured every 4-5 weeks. Cultures were maintained in the dark at 26±1°C. Embryogenic tissue was subcultured at 10-day intervals in liquid DCR medium. Embryogenic tissue of white spruce is not a callus as it consists of a mixture of highly vacuolated cells, single or small groups of densely cytoplasmic cells, and filamentous immature embryos consisting of a small group of apical, densely plasmatic cells attached to highly vacuolated, elongated suspensor cells (Joy *et al.*, 1991). Subculturing led to the proliferation of these various cells and tissues. Further development of somatic embryos was achieved by transferring these tissues onto AE medium with ABA (10-40 μM) as the sole phytohormone.

DEVELOPMENTAL STAGES IN ORGANIZED DEVELOPMENT

The use of model systems, which allows for reproducible reorganization with defined temporal stages, permits basic physiological and biochemical studies to be correlated with a specific developmental event. Thus, detailed histological and ultrastructural studies have been carried out with these four tissue culture systems.

1. Organogenesis *in vitro*

As pointed out by Hicks (1980), there are at least two patterns of organ formation *in vitro*. These are indirect organogenesis, in which the primary explant is induced to form callus prior to undergoing *de novo* organogenesis (e.g., the tobacco callus system) and direct organogenesis in which there is no intervening callus proliferation stage (e.g., radiata pine excised cotyledons). The process of *de novo* organogenesis in these two systems will be briefly outlined.

Histological studies with tobacco callus (Maeda and Thorpe, 1979; Ross *et al.*, 1973; Thorpe and Murashige, 1970) have shown that shoot primordium formation begins with the appearance of cell files on the base of the inoculum and some tracheary elements in the area between the cell files and the inoculum. By day 8 in culture, zones of preferential cell division activity are observed near the tracheary elements. In these zones which are mainly located in the lower half of the callus, meristemoids or meristem-like masses of cells arise beginning around day 9. These meristemoids become shoot primordia starting around day 12, with visible buds observable after day 14. Leafy vegetative shoots are clearly recognizable between days 18-21. Although the process is not synchronous, the key histological events leading to shoot primordium formation occur between days 6-14 in the dark-grown cultures.

In radiata pine, numerous mitotic figures can be seen throughout the cotyledons at the time of excision and all divisions are anticlinal. In the BA-free NSF cotyledons, cell division ceases by day 2 and cotyledonary maturation events occur (Yeung *et al.*, 1981). In contrast, in the SF cotyledons mitotic activity becomes localized in the epidermal and subepidermal cell layers in contact with the medium by day 3 (Villalobos *et al.*, 1985). Organized structures, term promeristemoids can be observed by day 5. Each of these 6-8 cell structures arises from a single subepidermal cell, which undergoes both periclinal and anticlinal divisions. These promeristemoids continue to develop, leading to the formation of meristemoids by day 10, which give the cotyledons a nodular appearance. Further development leads to formation of bud primordia by day 21. One interesting observation is that continued cytokinin and light are required for the development beyond the promeristemoid stage (Villalobos *et al.*, 1984b).

The key morphological feature of *de novo* organogenesis is the formation of the meristemoid. In both tobacco and radiata pine, these arise from vacuolated parenchyma cells (Ross *et al.*, 1973; Villalobos *et al.*, 1985). The cells of the meristemoid characteristically are small, isodiametric, and thin-walled with densely staining and prominent nuclei. In tobacco, the cells are densely plasmatic and microvacuolated, while in radiata pine the cells are vacuolated at the light microscope level and the dense cytoplasmic stage appears later. In both tissues, plasmodesmata are prominent between cells in the same (pro)meristemoid, but not between different meristemoids or other cells in the tissue. Based on these studies, as well as others with different systems, it appears that localized active cell divisions leading to a meristematic center or meristemoid is a general histological feature of *de novo* organogenesis.

2. Somatic Embryogenesis *in vitro*

In carrot, induction of embryogenic tissue occurs in the presence of auxin. Cell division is restricted to the more superficial region of the multicellular masses, while the central cells senesce and the meristematic layer dissociates into small densely cytoplasmic cell clusters (Wetherell, 1979). Groups of daughter cells often remain within the mother cell wall and thus are set apart from other groups. There are numerous plasmodesmata within each group of cells, which have thinner cell walls than the surrounding cells. Subculture of these cell masses in the presence of auxin leads to the proliferation of these cell clusters as well as the non-embryogenic cells within the tissue mass. On transfer to auxin-free medium, there is an initial acceleration of growth due to localized active multiplication of the densely cytoplasmic cells. These aggregates separate from the highly vacuolated cells, continue to divide and finally produce somatic embryos, which are released when they reach the globular stage. The process is comparable to zygotic embryogenesis, as structures similar to globular, heart and torpedo stage embryos can be found.

Using their single cell system, Komamine and his associates have been able to recognize four phases, referred to as 0, 1, 2, and 3, in carrot somatic embryogenesis (Fujimura and Komamine, 1980a; Komamine *et al.*, 1992). In phase 0, competent single cells (state 0) form embryogenic cell clusters (state 1) in the presence of auxin. During this phase, the cell clusters formed from single cells gain the ability to develop into embryos in the absence of auxin. During phase 1, cell clusters proliferate slowly and apparently without further differentiation. After phase 1, rapid cell division occurs in certain parts of the cell clusters, leading to the formation of globular embryos (phase 2). In phase 3, plantlets develop from globular embryos via heart-shaped and torpedo-shaped embryos.

The white spruce embryogenic tissue is white and mucilaginous in texture (Joy *et al.*, 1991). Among the highly vacuolated cells, single or small groups of densely cytoplasmic cells are present. Filamentous immature embryos consisting of a small group of apical, densely plasmatic cells attached to highly vacuolated, elongated suspensor cells are numerous. The apical cells of the filamentous embryos have large densely staining nuclei. Subculture of this tissue in the presence of auxin and cytokinin allows for the continued proliferation of this complex embryogenic tissue mass. For further development, the tissue is transferred to medium with ABA as the only phytohormone. Increased meristematic activity occurs within the densely plasmatic cells, particularly in the basal end of the head initially, so that by 3 weeks apical meristems, procambium and cotyledon primordia can be observed. The latter, usually about eight, can be seen as meristematic ridges. Further development of these regions and directional expansion of others lead to a recognizable mature, creamy- yellow embryo by week 4. At this stage, the suspensor begins to senesce and the embryo proper can be separated easily. A detailed, histological description of the development of white spruce somatic embryos has been given by Hakman *et al.* (1987).

PHYSIOLOGICAL ASPECTS OF ORGANIZED DEVELOPMENT

Appearance of an organ or an embryo is the endpoint of a series of developmental steps involving cellular dedifferentiation, acquisition of competence, cellular determination and only finally cellular differentiation (Christianson, 1987). While many studies have been concerned with optimizing the culture environment, few have dealt with the physiology and biochemistry of organized development. Furthermore, in most experimental systems, the major problems in trying to carry out basic studies on these morphogenic processes is lack of synchrony and the dilution effect (Thorpe, 1982). However, considerable progress in overcoming these problems has been made with carrot (see earlier, Nomura and Komamine, 1985).

What are the physiological requirements for initiating organized development? Concentration, diffusion or physiological gradients of materials from the medium into the tissue have been implicated in determining the loci at which primordium initiation begins in tobacco callus (Ross and Thorpe, 1973). This information was obtained by inverting the tissue at different times in culture, which altered the position at which organogenesis occurred. In addition, the timing of the SF process can be modulated by transfer of the tissues between permissive and non-permissive conditions, i.e., between SF and NSF media (Hammersley-Straw and Thorpe, 1987), and by the use of osmotic inhibition by high levels of sucrose (Hammersley-Straw and Thorpe, 1988). Also, the initiation of differentiation in the subepidermal cell layer of the surface of radiata pine cotyledons in contact with the medium (Yeung *et al.*, 1981; Villalobos *et al.*, 1985) offers support for the physiological gradient concept. However, this mechanism cannot explain how individual cells become activated and induced to undergo *de novo* organogenesis - or: a process which has at least three requirements, namely (a) cell dedifferentiation, (b) cell interaction, and (c) reaction to specific signals (Thorpe, 1980).

Physical and/or physiological isolation of a cell from its neighbours has been proposed as a prerequisite for organized development (Steward *et al.*, 1958). This idea was proposed on the basis of observations in carrot cell suspensions. While there is no evidence for physical isolation, based on ultrastructural studies with this system (Halperin and Wetherell, 1964), physiological isolation is possible (Thorpe, 1980). In morphogenic cell masses (meristemoids or embryogenic cell clusters), there is often a thick surrounding cell wall (e.g., Kohlenbach, 1977), and a virtual absence of plasmodesmata between these and the surrounding cells/tissues; while within the masses there are thinner cell walls and a high concentration of plasmodesmata, indicating possible cellular communication.

An alternative view although not mutually exclusive, which emerged from the studies of Skoog and his associates, as mentioned earlier, is that organized development arises through quantitative interactions between diverse growth factors including phytohormones and metabolites (Skoog and Miller, 1957). Precise balances of substances are placed in the medium to allow for precise balances within the tissues or cell masses for organized development. This is particularly true of the concentration and type of inorganic salts, especially N in the form of NO_3^- or NH_4^+, and phytohormones, particularly auxin and cytokinin (Thorpe, 1980; 1988a). Indeed the control of organogenesis by endogenous auxin/cytokinin balances has been demonstrated through the use of disarmed Ti plasmids of *Agrobacterium tumefaciens* (see Thompson and Thorpe, 1990).

Since phytohormones play such an important role in organized development (Chandler and Thorpe, 1986; see Thorpe, 1980), we have carried out experiments looking directly and indirectly at the exogenous effects as well as endogenous contents and metabolism of several phytohormones during organogenesis and to a lesser extent somatic embryogenesis.

Indirect studies based on changes in cathodic isoperoxidases in tobacco callus and epidermal explants have indicated that endogenous auxin content was modulated in favour of the optimum auxin/cytokinin ratio for shoot and root initiation (Thorpe and Gaspar, 1978; Thorpe *et al.*, 1978). Tobacco callus contains GA-like substances and is capable of metabolizing exogenous GAs (Lance *et al.*, 1976a,b). The level and the spectrum of endogenous GAs change during shoot formation. This suggests that GAs are involved in normal tissue growth and differentiation, but that tobacco callus synthesizes enough for the organogenetic process. In excised radiata pine cotyledons, the contents of endogenous IAA and ABA also change during bud induction (Macey, Reid and Thorpe, unpubl. results). Lastly, we have examined the role of C_2H_4 and CO_2 in both organogenic systems. Ethylene which accumulated in the headspace of shoot-forming tobacco callus inhibited organogenesis early in culture but it speeded up primordium formation later in culture (Huxter *et al.*, 1981). In radiata pine, the accumulation of both C_2H_4 and CO_2 early in the culture period promoted shoot formation, while excessive accumulation after bud initiation caused some dedifferentiation (Kumar *et al.*, 1987). Elimination of these gases inhibited bud formation, but O_2 was needed for their promotive effects, presumably by maintaining oxidative metabolism. The effects of exogenous phytohormones such as gibberellins, ABA, various auxins, etc., mainly on shoot formation in tobacco callus and/or excised cotyledons of radiata pine were examined (see Thorpe, 1980; 1988b). In general the application of growth regulators, other than auxin (in tobacco) and cytokinin, repressed organogenesis, and in the case of radiata pine cotyledons enhanced callus formation (Biondi and Thorpe, 1982), although the addition of ABA partially reversed the gibberellin (GA)-repression of shoot formation in tobacco callus (Thorpe and Meier, 1973).

Unfortunately, most of the studies carried out on the physiology and biochemistry of somatic embryogenesis in carrot deal with embryo development rather than the whole process, since the usual approach has been to transfer embryogenic clusters to auxin-free medium (Thorpe, 1982, 1988a). Addition of auxin or anti-auxins to embryogenic clusters inhibited embryo development (Fujimura and Komamine, 1975) although endogenous auxin is involved in embryo development (Fujimura and Komamine, 1979). Different auxins had different effects on embryo formation (Chandra *et al.*, 1978; Kamada and Harada, 1979a,b), but several aryloxyalkanecarboxylic acids enhanced embryo formation (Chandra *et al.*, 1978). In contrast, cytokinins generally, even at low concentrations, repressed somatic embryogenesis in carrot (Kamada and Harada, 1979a); the exception being zeatin at low concentrations (Fujimura and Komamine, 1980; Nomura and Komamine, 1985). Ethylene (and ethephon) (Tisserat and Murashige, 1977) and gibberellic acid (Fujimura and Komamine, 1975; Kamada and Harada, 1979a) suppressed embryo formation. In carrot, the spectrum and amount of endogenous GAs varied between embryogenic and non- embryogenic tissues plus/minus auxin (Noma *et al.*, 1982), suggesting a role for endogenous GAs in somatic embryogenesis. Lastly, while exogenous ABA has been shown to play a role in enhancing normal somatic embryogenesis in several species including white spruce (Hakman and von Arnold, 1988), it inhibits the process in carrot (Fujimura and Komamine, 1975), but changes in endogenous ABA with culture development were observed (Kamada and Harada, 1981).

Inorganic N in the form of NH_4^+ is required for the initiation of embryogenesis in carrot cell culture, since NO_3^- alone is insufficient (Halperin and Wetherell, 1965); the addition of as little as 0.1 mM NH_4Cl to the medium allowed for embryo formation (Wetherell and Dougall, 1976). We have recently found, using ^{15}N NMR, that NH_4^+ was taken up from the medium earlier and utilized faster than NO_3^- in both carrot and white spruce embryogenic tissues (Joy, Vogel and Thorpe, unpubl.). In addition to these qualitative effects of inorganic N, no embryos were formed in carrot at low levels of total N (Reinert *et al.*, 1967). Addition of N compounds, particularly NH_4NO_3 and glutamine enhanced embryogenesis (Reinert and Tazawa, 1969). Several amino acids added to medium containing 20 mM KNO_3 stimulated embryogenesis, alanine being particularly effective (Kamada and Harada, 1979b). Alanine was found to be rapidly transformed to glutamate via alanine aminotransferase and utilized as a N source (Kamada and Harada, 1984). There is also an apparent specific requirement for K^+ (Brown *et al.*, 1976).

Several other factors which influence the physiology of cultured cells also have a positive affect on embryogenesis. These include density of cells in culture (Halperin, 1967; Hari, 1980) with a high cell density being required for cell cluster formation (Nomura and Komamine, 1985) and a lower cell density for embryo development from embryogenic cells (Fujimura and Komamine, 1979). Another factor is conditioned medium (Hari, 1980). Although the factors involved in conditional media are unknown, the presence of small molecular weight secreted proteins in the medium has been correlated with early stages of somatic embryo development (De Vries *et al.*, 1988, and this volume). Osmotic stress also can enhance embryogenesis (Kamada *et al.*, 1986). Lastly, dissolved O_2 below critical levels has been shown to favor differentiation of carrot embryos from cell clusters (Kessel and Carr, 1972). Under these conditions, CN-sensitive respiration and cellular concentrations of ATP increased (Kessel *et al.*, 1977). The ATP level could also be increased by adding adenosine to the medium, which also enhanced embryo formation. However, starting with single cells of carrot, high levels of O_2 (40%) in the presence of auxin led to the highest number of embryos on transfer to auxin-free medium (Nomura and Komamine, 1985). Endogenous currents have been found to traverse somatic embryos of carrot, entering the apical pole and leaving the region of the presumptive radicle in globular embryos (Brawley *et al.*, 1984). Exogenous IAA reversibly inhibited these currents. Such electrical polarity was also found in embryogenic cell clusters in the presence of 2,4-D (Gorst *et al.*, 1987). It appears that the inward current is due largely to K^+ influx and the outward current to active H^+ efflux (Rathore *et al.*, 1988).

BIOCHEMICAL ASPECTS OF ORGANOGENESIS

We have pursued this topic using both tobacco callus and excised cotyledons of radiata pine as experimental material. Early studies using histo- and cyto-chemical approaches revealed that cells in shoot-forming regions of tobacco callus stained more intensely for RNA and protein but not DNA (Thorpe and Murashige, 1970). Starch accumulated in these SF tissues and declined during meristemoid formation. In radiata pine, increased histochemical staining for both DNA and RNA as well as protein was observed in the shoot-forming regions (Patel and Thorpe, 1984). These findings were substantiated through an autoradiographic examination of the incorporation of label from thymidine, uridine and leucine in SF and NSF cotyledons (Villalobos *et al.*, 1984a). This latter study showed that the accumulation of the macromolecules was due to synthesis and not reduced degradation, and differences between SF and NSF cotyledons were evident after 24 hrs in culture. The protein accumulation was reflected in changes in specific enzymes (Patel and Thorpe, 1984) and included more intense staining for succinic dehydrogenase, peroxidase, acid phosphatase and ATPase in SF regions. Thus, it is apparent that organogenesis involves a switch in development, reflected at the level of macromolecular synthesis and accumulation, as well as increases in specific enzyme activities, directly in those cells destined to give rise to shoots (Thompson and Thorpe, 1990).

Further studies on the accumulation and utilization of starch during meristemoid formation established that the SF process is a highly energetic one (Thorpe, 1980). Starch was synthesized throughout the culture period, but its decline was due to enhanced rates of degradation (Thorpe and Meier, 1974). Interestingly, newly synthesized starch was not immediately degraded (Thorpe *et al.*, 1986). The breakdown of starch, along with free sugars from the medium (Thorpe 1974), provided substrates for increased respiration in the SF part of the callus compared to NSF tissue as measured directly (Thorpe and Meier, 1972) and through enhanced activities of the glycolytic and pentose phosphate pathways and increases in the rate and pattern of glucose oxidation (Thorpe and Laishley, 1973). Part of

1980a). The rate of rRNA was lower during embryo development than in the presence of auxin, while the reverse was observed for the rate of poly(A)$^+$ RNA synthesis (Sengupta and Raghavan, 1980b). Rates of protein synthesis also increased after transfer to auxin-free medium.

Starch accumulation, a prominent feature of embryogenic cell clusters of carrot, disappeared during embryo formation (Street and Withers, 1974). An increase in lipid synthesis occurred during the globular and heart stages of somatic embryogenesis in carrot (Warren and Fowler, 1979). Although quantitative rather than qualitative changes in fatty acid content were observed, globular stage embryos possessed some long chain (>20) fatty acids not found in earlier or later stages of embryo development. In white spruce, storage product deposition was followed during embryo development from the filamentous immature to the mature cotyledonary stage (Joy *et al.*, 1991). During early stages mostly starch was produced, whereas during later stages starch, lipids and proteins were stored. Quantitatively, cotyledonary stage somatic embryos had less lipid and protein and more starch than equivalent stage zygotic embryos, indicating that the somatic embryos were not identical to the zygotic ones.

In both carrot and white spruce, resonances arising through the feeding of $^{15}N\text{-}NH_4^+$ and NO_3^- showed that qualitative and quantitative changes occurred in N assimilation (Joy, Vogel and Thorpe, unpublished.). Inorganic N initially was incorporated principally into glutamine, glutamate and alanine, and with embryo development into arginine and aliphatic amines as well. These studies showed that during embryo development, the flow of N occurred via the GS/GOGAT (glutamine synthetase/glutamate synthase) pathway and subsequently into the ornithine cycle and finally into polyamines, urea and GABA (γ-amino butyric acid).

Earlier studies had shown no significant differences in endogenous amino acids in carrot during embryo development (Kamada and Harada, 1984), but the involvement of polyamines was suggested, since higher polyamine levels was found during embryo development (Fienberg *et al.*, 1984; Montague *et al.*, 1978; also see Bagni *et al.*; and Tiburcio *et al.*, this volume). Also, higher levels of arginine decarboxylase was found in embryogenic cell clusters than non-embryogenic cells (Montague *et al.*, 1979). As well, ornithine carbamoyltransferase increased during embryo development (Baker *et al.*, 1983). Auxin suppressed the activities of both arginine carboxylase and S-adenosylmethionine (SAM) decarboxylase (Fienberg *et al.*, 1984). An inhibitor of this last enzyme reduced polyamine levels and enhanced the cellular levels of 1-aminocyclopropane-1-carboxylic acid, a precursor of C_2H_4. These results suggest the regulation of competition between C_2H_4 and polyamine biosynthesis for SAM, and the subsequent direction of organized development. These studies also indicate that a shift in metabolism occurs during somatic embryogenesis.

MOLECULAR ASPECTS OF ORGANIZED DEVELOPMENT

Organogenesis is the final expression of a number of stages in culture. Within a particular tissue, certain cells possess the ability to recognize and respond to specific signals (Thorpe, 1980). Similarly, only one cell type in carrot has the capacity to become determined for embryogenesis (Nomura and Komamine, 1985). Nevertheless, these cells need the appropriate phytohormonal and chemical stimuli to differentiate. The shifts in metabolism which precede and are coincident with organized development, as outlined above, are indicative but not proof that differentiation is occurring at the molecular level (Thompson and Thorpe, 1990). Thus an analysis of gene expression during organized development is needed. With improvements in the methods of molecular analysis, slow progress is being made in this direction (e.g., see Thompson and Thorpe, 1990; Komamine *et al.*, 1992; De Vries *et al.*, and Hatzopoulos this volume).

Attempts have been made to find molecular markers of somatic embryogenesis. For example, in carrot 2D gel electrophoresis revealed two embryogenic proteins in somatic embryos that were undetectable in undifferentiated cells (Sung and Okimoto, 1981). Further studies, e.g., Hatzopoulos *et al.* (1990) showed changes in specific polypeptides during embryo development and examined their regulation (see Hatzopoulos, this volume). Another study showed that of the 400 translated proteins, only two polypeptides appeared and two disappeared in developing embryos prior to the globular stage (Fujimura and Komamine, 1982). More recent studies revealed the presence of several markers, namely polypeptides, mRNAs, antigens against monoclonal antibodies, which can be detected during the expression of totipotency in carrot cells, but they disappear during its loss (Komamine *et al.*, 1992). Four organ-specific genes have been isolated from hypocotyls and roots by differential screening. They were expressed preferentially after the globular-heart stages of embryogenesis and were strongly suppressed by auxin. One gene, isolated by differential screening of embryogenic cell clusters, was expressed strongly and transiently during the proglobular and globular stages.

During organogenesis in excised cotyledons of radiata pine 54 polypeptides whose synthesis and accumulation were affected by the culture conditions have been identified (Thompson and Thorpe, unpublished.). These could be divided into seven classes based on their pattern of appearance. The primary class of interest consisted of three polypeptides whose synthesis and/or accumulation was enhanced in, but not unique to, SF cotyledons. The second and largest class of interest consisted of 29 polypeptides whose synthesis appeared to be specifically repressed in SF cotyledons, suggesting that part of the role of cytokinin (CK) in inducing shoot formation lay in repressing maturation-related changes in the SF cotyledons. This interpretation can help to explain the absolute requirement for CK during the first 3 days in culture (Biondi and Thorpe, 1982; Villalobos *et al.*, 1984a). The remaining five classes could not be directly related to the organogenic capacity of the cotyledons.

At least part of the problem in using 2D-gels for analysis of morphogenic-specific events, is the fact that the technique can only resolve about 1000 polypeptides, in contrast to the production of at least 20,000 mRNA species in any plant organ (Kamalay and Goldberg, 1980). Furthermore, the polypeptides resolved are those that are most abundant while the morphogenic-specific ones, being regulatory in nature, are of low abundance. Also in studying gene expression during embryogenesis, in systems such as carrot or white spruce, the embryogenic tissues induced in the presence of auxin are already producing embryogenic-specific polypeptides (see Hatzopoulos, this volume). Thus other approaches, including cDNA probing, immunoblotting, anti-sense RNA technology, etc., are needed and are being used. A further problem is that of being able to ascribe a specific morphogenic role to phytohormone-induced changes in gene expression (Thompson and Thorpe, 1990). In spite of these difficulties, great progress is being made in applying these approaches to analyzing the process of organized development, although there is a large gap between identifying stage-specific genes and assigning functions to them.

CONCLUDING THOUGHTS

The material presented above clearly indicates that *de novo* organized development, whether organogenesis or embryogenesis, whether in a herbaceous angiosperm or a gymnosperm, has certain features in common at the physiological, biochemical and structural levels. Even so, at present we are not close to answering such fundamental questions, as "what activates a single cell so that it will respond to signals and undergo organized development?" or "why do some activated cells produce embryos and others only half embryos, i.e., either shoot or root primordia?" In examining the physiological and biochemical aspects of organized development with the most widely used experimental systems, it is clear that most studies have been fragmented. Nevertheless, these studies have begun to describe the consequences of the selective gene activity which results in organized development. In all cases, localized active cell division leading to a meristemoid or embryogenic cluster is the route of reorganization. The process has a high energy requirement and is metabolically very active. Polymeric reserves such as starch and lipid are mobilized during the process. Carbon and nitrogen metabolism interact in meeting the needs of the enhanced metabolism. Nucleic acid, amino acid and protein metabolism are centrally involved in the differentiation process. Phytohormones play a pivotal but as yet not fully understood role in the activation process. While none of these findings individually or even collectively are surprising, together they are allowing for a clearer understanding of the differentiation process. Coupled with molecular approaches, one can confidently expect the type of information that will dramatically improve our understanding of *de novo* organized development, the process which best exemplifies the totipotency of the plant cell.

ACKNOWLEDGEMENTS

The author acknowledges with gratitude the contributions of colleagues, graduate students, post-doctoral fellows, visiting scientists and research assistants to the personal research reported here. In particular he thanks recent graduate students, Richard W. Joy IV, Prakash P. Kumar, Donna E. Macey and Michael R. Thompson, and colleagues David Reid and Hans Vogel, who have contributed as yet unpublished material. The author also thanks his secretary, Josie Cleland, for typing the ms and for help in organizing the references. Financial support from the Natural Sciences and Engineering Research Council of Canada and the University of Calgary also is recognized gratefully.

REFERENCES

Aitken J, Horgan KJ, Thorpe TA (1981) Influence of explant selection on the shoot-forming capacity of juvenile tissue of *Pinus radiata*. *Can J For Res* 11: 112-117.

Aitken Christie J, Singh AP, Horgan KJ, Thorpe TA (1985) Explant developmental state and shoot formation in *Pinus radiata* cotyledons. *Bot Gaz* 146: 196-203.

Ammirato PV (1984) Induction, maintenance, and manipulation of development in embryogenic cell suspension cultures. *In*: Cell Culture and Somatic Cell Genetics of Plants, vol 1, Vasil IK (ed), Academic Press, New York, pp 130-151.

Ashihara H, Fujimura T, Komamine A (1981) Pyrimidine nucleotide biosynthesis during somatic embryogenesis in a carrot cell suspension culture. *Z. Pflanzenphysiol* 104: 129-137.

Baker SR, Jones LH, Yon RJ (1983) Ornithine carbamoyl transferase activity and embryogenesis in a carrot cell suspension culture. *Phytochemistry* 22: 2167-2169.

Beaudoin-Eagan LD, Thorpe TA (1983) Shikimate pathway activity during shoot initiation in tobacco callus cultures. *Plant Physiol* 73: 228-232.

Beaudoin-Eagan LD, Thorpe TA (1984) Turnover of shikimate pathway metabolites during shoot initiation in tobacco callus cultures. *Plant Cell Physiol* 25: 913-921.

Beaudoin-Eagan LD, Thorpe TA (1985a) Shikimate pathway enhancement in the lower half of shoot-forming tobacco callus. *J Plant Physiol* 120: 87-90.

Beaudoin-Eagan LD, Thorpe TA (1985b) Tyrosine and phenylalanine ammonia lyase activities during shoot initiation in tobacco callus cultures. *Plant Physiol* 78: 438-441.

Bender L, Joy IV RW, Thorpe TA (1987a) Studies on [^{14}C]-glucose metabolism during shoot bud induction in cultured cotyledon explants of *Pinus radiata* D. *Don*. *Physiol Plant* 69: 428-434.

Bender L, Joy IV RW, Thorpe TA (1987b) [^{14}C]-Glucose metabolism during shoot bud development in cultured cotyledon explants of *Pinus radiata*. *Plant Cell Physiol* 28: 1335-1338.

Biondi S, Thorpe TA (1982) Growth regulator effects, metabolite changes and respiration during shoot initiation in cultured cotyledon explants of *Pinus radiata*. *Bot Gaz* 143: 20-25.

Biondi S, Torrigiani P, Sansovini A, Bagni N (1988) Inhibition of polyamine biosynthesis by dicyclohexamine in cultured cotyledons of *Pinus radiatia*. *Physiol Plant* 72: 471-476.

Brawley SH, Wetherell DF, Robinson KR (1984) Electrical polarity in embryos of wild carrot precedes cotyledon differentiation. *Proc Natl Acad Sci USA* 81: 6064-6067.

Brown DCW, Thorpe TA (1980a) Adenosine phosphate and nicotinamide adenine dinucleotide pool sizes during shoot initiation in tobacco callus. *Plant Physiol* 65: 587-590.

Brown DCW, Thorpe TA (1980b) Changes in water potential and its components during shoot formation in tobacco callus. *Physiol Plant* 49: 83-87.

Brown DCW, Thorpe TA (1982) Mitochondrial activity during shoot formation and growth in tobacco callus. *Physiol Plant* 54: 125-130.

Brown DCW, Thorpe TA (1986) Plant regeneration by organogenesis. *In:* Cell Culture and Somatic Cell Genetics in Plants, vol. III: Plant Regeneration and Genetic Variability, Vasil IK (ed), Academic Press, New York, pp 49-66.

Brown DCW, Leung DWM, Thorpe TA (1979) Osmotic requirement for shoot formation in tobacco callus. *Physiol Plant* 46: 36-41.

Brown S, Wetherell DF, Dougall DK (1976) The potassium requirement for growth and embryogenesis in wild carrot suspension cultures. *Physiol Plant* 37: 73-79.

Lu C-Y, Thorpe TA (1987) Somatic embryogenesis and plantlet regeneration in cultured immature embryos of *Picea glauca*. *J Plant Physiol* 128: 297- 302.

Maeda E, Thorpe TA (1979) Shoot histogenesis in tobacco callus cultures. *In Vitro* 15: 415-424.

Montague MJ, Koppenbrink JW, Jaworski EG (1978) Polyamine metabolism in embryogenic cells of *Daucus carota*. I. Changes in intracellular content and rates of syntheses. *Plant Physiol* 62: 430-433.

Montague MJ, Armstrong TA, Jaworski EG (1979) Polyamine metabolism in embryogenic cells of *Daucus carota*. II. Changes in arginine decarboxylase activity. *Plant Physiol* 63: 341-345.

Murashige T (1961) Suppression of shoot formation in cultured tobacco cells by gibberellic acid. *Science* 134: 280.

Murashige T, Skoog F (1962) A revised medium for rapid growth and bioassays with tobacco tissue cultures. *Physiol Plant* 15: 473-497.

Nobécourt P (1939) Sur les radicelles naissant des cultures de tissus du tubercule de carotte. *CR Seances Soc Biol Ses Fil* 130: 1271-1272.

Noma M, Huber J, Ernst D, Pharis RP (1982) Quantitation of giberellins and the metabolism of [^{3}H]gibberellin A_1 during somatic embryogenesis in carrot and anise cell cultures. *Planta* 155: 369-376.

Nomura K, Komamine A (1985) Identification and isolation of single cells that produce somatic embryos at a high frequency in a carrot suspension culture. *Plant Physiol* 79: 988-991.

Nomura K, Komamine A (1986a) Polarized DNA synthesis and cell division in cell clusters during somatic embryogenesis from single carrot cells. *New Phytol* 104: 25-32.

Nomura K, Komamine A (1986b) Somatic embryogenesis in carrot cells. *Dev Growth Differ* 28: 511-517.

Obata-Sasamoto H, Villalobos VM, Thorpe TA (1984) ^{14}C-metabolism in cultured cotyledon explants of radiata pine. *Physiol Plant* 61: 490-496.

Patel KR, Thorpe TA (1984) Histochemical examination of shoot formation in cultured cotyledon explants of radiata pine. *Bot Gaz* 145: 312-322.

Plumb-Dhindsa PL, Dhindsa RS, Thorpe TA (1979) Non-autotrophic CO_2 fixation during shoot formation in tobacco callus. *J Exp Bot* 30: 759-767.

Rathore KS, Hodges TK, Robinson KR (1988) Ionic basis of currents in somatic embryos of *Daucus carota*. *Planta* 175: 280-289.

Reilly KJ, Washer J (1977) Vegetative propagation of radiata pine by tissue culture: plantlet formation from embryonic tissue. *NZJ For Sci* 7: 199- 206.

Reinert J (1958) Morphogenese und ihre kontrolle an gewebekulturen aux karotten. *Naturwissenschaften* 45: 344-345.

Reinert J, Tazawa M (1969) Wirkung von strickstoffverbindungen und von auxin auf die embryogenese in gewebekulturen. *Planta* 87: 239-248.

Reinert J, Tazawa M, Semenoff S (1967) Nitrogen compounds as factors of embryogenesis *in vitro*. *Nature* 216: 1215-1216.

Ross MK, Thorpe TA (1973) Physiological gradients and shoot initiation in tobacco callus cultures. *Plant Cell Physiol* 14: 473-480.

Ross MK, Thorpe TA, Costerton JW (1973) Ultrastructural aspects of shoot initiation in tobacco callus cultures. *Amer J Bot* 60: 788-795.

Sengupta C, Raghavan V (1980a) Somatic embryogenesis in carrot cell suspension. I. Pattern of protein and nucleic acid synthesis. *J Exp Bot* 31: 247-258.

Sengupta C, Raghavan V (1980b) Somatic embryogenesis in carrot cell suspension. II. Synthesis of ribosomal and poly(A)$^+$ RNA. *J Exp Bot* 31: 259-268.

Skoog F (1944) Growth and organ formation in tobacco tissue cultures. *Am J Bot* 31: 19-24.

Skoog F, Miller CO (1957) Chemical regulation of growth and organ formation in plant tissues cultured *in vitro*. *Symp Soc Exp Biol* 11: 118-131.

Steward FC, Shantz EM (1956) - The chemical induction of growth in plant tissue cultures. *In:* The Chemistry and Mode of Action of Plant Growth Substances, Wain RL, Wightman FC (eds), Academic Press, New York, pp 165-187.

Steward CF, Mapes MO, Mears K (1958) Growth and organized development of cultured cells. II. Organization in cultures grown freely from suspended cells. *Am J Bot* 45: 705-709.

Street HE, Withers LA (1974) The anatomy of embryogenesis in culture. *In:* Tissue Culture and Plant Science, Street HE (ed), Academic Press, London, pp 71-100.

Sung ZR, Okimoto R (1981) Embryonic proteins in somatic embryos of carrot. *Proc Natl Acad Sci USA* 78:3683-3687.

Thompson MR, Thorpe TA (1990) Biochemical perspectives in tissue culture for crop improvement. *In:* Biochemical Aspects of Crop Improvement, Khanna KR (ed), CRC Press, Boca Raton, FL, pp 328-358.

Thorpe TA (1974) Carbohydrate availability and shoot formation in tobacco callus cultures. *Physiol Plant* 30: 77-81.

Thorpe TA (1980) Organogenesis *in vitro*: Structural, physiological and biochemical aspects. *Int Rev Cytol Suppl* 11A: 71-111.

Thorpe TA (1982) Callus organization and *de novo* formation of shoots, roots and embryos *in vitro*. *In:* Techniques and Applications of Plant Cell and Tissue Culture to Agriculture and Industry, Tomes D, Ellis BE, Harney PM, Kasha KJ, Peterson RL (eds), University of Guelph, Ontario, pp 115-138.

Thorpe TA (1983) Morphogenesis and regeneration in tissue culture. *In*: Genetic Engineering: Applications to Agriculture, Beltsville Symp No 7, Owens LD (ed), Rowan and Allanheld, Totowa NJ, pp 258- 303.

Thorpe TA (1988a) *In vitro* somatic embryogenesis. *ISI Atlas of Science - Animal and Plant Sciences* 1: 81-88.

Thorpe TA (1988b) Physiology of bud induction in conifers *in vitro*. *In:* Genetic Manipulation of Woody Plants, Hanover JW, Keathley DE (eds), Plenum Press, New York, pp 167-184.

Thorpe TA (1990) The current status of plant tissue culture. *In:* Plant Tissue Culture: Applications and Limitations, Bhojwani SS (ed), Elsevier Science Publ, Amsterdam, pp 1-33.

Thorpe TA, Beaudoin-Eagan LD (1984) ^{14}C-metabolism during growth and shoot formation in tobacco callus. *Z Pflanzenphysiol* 113: 337-346.

Thorpe TA, Gaspar T (1978) Changes in isoperoxidases during shoot formation in tobacco callus. *In Vitro* 14: 522-526.

Thorpe TA, Laishley EJ (1973) Glucose oxidation during shoot initiation in tobacco callus cultures. *J Exp Bot* 24: 1082-1089.

Thorpe TA, Meier DD (1972) Starch metabolism, respiration and shoot formation in tobacco callus cultures. *Physiol Plant* 27: 365-369.

Thorpe TA, Meier DD (1973) Effects of gibberellic acid and abscisic acid on shoot formation in tobacco callus cultures. *Physiol Plant* 29: 121-124.

Thorpe TA, Meier DD (1974) Enzymes of starch metabolism in *Nicotiana tabacum* callus. *Phytochem* 13: 1329-1334.

Thorpe TA, Murashige T (1970) Some histochemical changes underlying shoot initiation in tobacco callus cultures. *Can J Bot* 48: 277-285.

Thorpe TA, Patel KR (1984) Clonal propagation: Adventitious buds. *In:* Cell Culture and Somatic Cell Genetics of Plants, Vasil IK (ed), vol 1, Academic Press, New York, pp 49-60.

Thorpe TA, Joy IV RW, Leung DWM (1986) Starch turnover in shoot-forming tobacco callus. *Physiol Plant* 66: 58-62.

Thorpe TA, Tran Thanh Van K, Gaspar T (1978) Isoperoxidases in epidermal layers of tobacco and changes during organ formation *in vitro*. *Physiol Plant* 44: 388-394.

Tisserat B, Murashige T (1977) Repression of asexual embryogenesis *in vitro* by some plant growth regulators. *In Vitro* 13: 799-805.

Tisserat B, Esan EB, Murashige T (1979) Somatic embryogenesis in angiosperms. *Hortic Rev* 1: 1-78.

Tran Thanh Van K (1980) Control of morphogenesis by inherent and exogenously applied factors in thin cell layers. *Int Rev Cytol Suppl* 11A: 175-194.

Villalobos VM, Oliver MJ, Yeung EC, Thorpe TA (1984a) Cytokinin-induced switch in development in excised cotyledons of radiata pine culture *in vitro*. *Physiol Plant* 61: 483-489.

Villalobos VM, Leung DWM, Thorpe TA (1984b) Light-cytokinin interaction in shoot formation in cultured cotyledon explants of radiata pine. *Physiol Plant* 61: 497-504.

Villalobos VM, Yeung EC, Thorpe TA (1985) Origin of adventitious shoots in excised radiata pine cotyledons cultured *in vitro*. *Can J Bot* 63: 2172- 2176.

von Arnold S, Eriksson T (1981) *In vitro* studies of adventitious shoot formation in *Pinus contorta*. *Can J Bot* 59: 870-874.

Warren GS, Fowler MW (1979) Changing fatty acid composition during somatic embryogenesis in cultures of *Daucus carota*. *Planta* 144: 451-454.

Wetherell DF (1979) *In vitro* embryoid formation in cells derived from somatic plant tissues. *In*: Propagation of Higher Plants through Tissue Culture - A Bridge Between Research and Application, Hughes KW, Henke R, Constantin M (eds), Conf-780411, US Tech Inf Serv, Springfield, VA, pp 102-124.

Wetherell DF, Dougall DK (1976) Sources of nitrogen supporting growth and embryogenesis in cultured wild carrot tissue. *Physiol Plant* 37: 97-103.

White PR (1939) Controlled differentiation in plant tissue culture. *Bull Torrey Bot Club* 66: 507-513.

White PR (1943) Nutrient deficiency studies and an improved inorganic nutrient for cultivation of excised toamato roots. *Growth* 7: 53-65.

Yeung EC, Aitken J, Biondi S, Thorpe TA (1981) Shoot histogenesis in cotyledon explants of radiata pine. *Bot Gaz* 142: 494-501.

RELATION BETWEEN SOME CYTOLOGICAL, BIOCHEMICAL, MOLECULAR MARKERS AND PLANT MORPHOGENESIS

Kiêm Tran Thanh Van and **Cyrille Amin Gendy**

Laboratoire de Physiologie Végétale
Centre National de la Rechèrche Scientifique
C.N.R.S, 91190, Gif sur Yvette, FRANCE

INTRODUCTION

In vivo and *in vitro* genesis of organs and functions in plants is modulated by multiple factors including environmental factors inputs, exogenous supply and/or endogenous variation of plant growth regulators. Auxins, cytokinins, gibberellins, ethylene and abscisic acid have long been recognized as acting on *in vivo* plant growth and development and on *in vitro* morphogenesis such as organogenesis, embryogenesis and callus formation. Other classes of molecules such as polyamines, lectins, flavonoids, oligosaccharides (Fry, 1986; Lerouge *et al.*, 1990; Tran Thanh Van *et al.*, 1985) and salicilic acid (Raskin, 1992), have been quoted in the literature as acting on plant and cell growth. More recently, glycoproteins have been shown to be involved in somatic embryogenesis (De Jong *et al.*, 1992; LoSchiavo *et al.*, 1990) and lectins in legume/rhizobium recognition (Diaz *et al.*, 1989).

Some of these molecules were once thought to play a "hormonal role". This concept, based only on the biological activity obtained with relatively limited doses and on the transport of these molecules within the plant, has been widely accepted although strong evidence has not yet been obtained. From this concept, copied from the animal field, numerous efforts were developed towards the isolation of receptors or at least of specific binding proteins, to the search for signal transduction pathways and for specific gene(s) induction. Although followed for decades, these efforts have so far not yielded clear-cut data, due to the lack of specific responses to specific signal molecules in plant systems. The absence of cell migration during embryogenesis in the plant system is, however, an advantage for the study of morphogenic differentiation. In this chapter, we review a few models of morphogenesis in which cytological, biochemical and molecular changes can be paralled or correlated.

Markers of morphogenic differentiation can be identified when mutant lines are available as in *Arabidopsis*. However, there are more mutants in the shape and the localization of

Morphogenesis in Plants, Edited by K.A. Roubelakis-Angelakis
and K. Tran Thanh Van, Plenum Press, New York, 1993

conditioned by the appropriate stage of wild type embryos. As for the reversion of inhibition by tunicamycin mentioned above, the reversion effect here was also heat sensitive and protease sensitive. The active component was found to be an excreted protein of 32 kD, a glycosylated acidic endochitinase. The 38 kD protein with peroxidase activity had no effect. The embryos which were developed after the addition of the endochitinase formed a normal epidermis. This suggests that, as well as for the peroxidase mentioned above, the subtrates must be easily accessible to the excreted proteins and therefore could be localized in the cell wall of the outer cells of the "nodular structures" which differentiated an epidermal layer during the later globular stage.

This set of results demonstrated that proteins excreted at different developmental stages play specific roles in somatic embryogenesis, and defined cytological traits were closely related to specific protein (highly vacuolated cells in the absence of the 38 kD protein, well structured epidermis in *ts 11* mutant lines in the presence of 32 kD protein). Therefore, a balance between the rythm of cell division, the orientation of the cell plate and the intensity of cell expansion is likely to be the key to monitor somatic embryogenesis.

This observation holds true for other patterns of morphogenesis. One can draw from these results the importance of the structure and properties of the cell wall and the function(s) of its component parts in the building of plant organs. Specific signals for cell-cell interaction are needed in order to coordinate the spatial and temporal organisation of the cells into globular shaped proembryos or into a specific organ pattern distinct from irregular clumps of cells devoid of an outer epidermis layer. Therefore, cell cohesion as well as a coordinated orientation of the cell division plate are important in order to give rise to an epidermal layer with a core of meristematic cells. Plant cell cohesion depends in part upon the composition of the cell wall middle lamella (in particular on the activity of cell wall hydrolases) and also on specific signalling factors. One of the candidate types of molecules for cell signalling are special glycoproteins or lectins. In *Dictyostelium discoidum*, these molecules are shown to be more actively synthesized in the aggregated cells of the "slug" stage than the ameboid cells (Rosen *et al.*, 1973). We have shown, in *Psophocarpus tetragonolobus*, that there is a correlation between the state of differentiation (i.e. friable callus cells versus cells initiating organogenesis) and the level of intracellular lectins on one hand and of those lectins secreted into the culture medium on the other (Meimeth *et al.*, 1982).

In summary, for somatic embryogenesis (which represents a relatively simple model of morphogenic differentiation) and for organogenesis, one can hypothesize that chemical compounds such as auxins, ethylene, cytokinins, ions and/or pH, may play multiple roles and in particular they may modify cell wall structures and properties via induction or activation of specific cell wall enzymes (Gendy, 1991; Tran Thanh Van *et al.*, 1985). Furthermore, glycoproteins (some of which are lectins) and cell wall oligosaccharides could be involved in the cell-cell interaction.

CONTROLLED ORGANOGENESIS

The tobacco thin cell layer system, which offers the possibility to program a variety of different patterns of morphogenesis, has been presented in several reviews (Cousson *et al.*, 1992; Tran Thanh Van, 1973a, 1980, 1991; Tran Thanh Van *et al.*, 1990). Flowers (Figs 1,

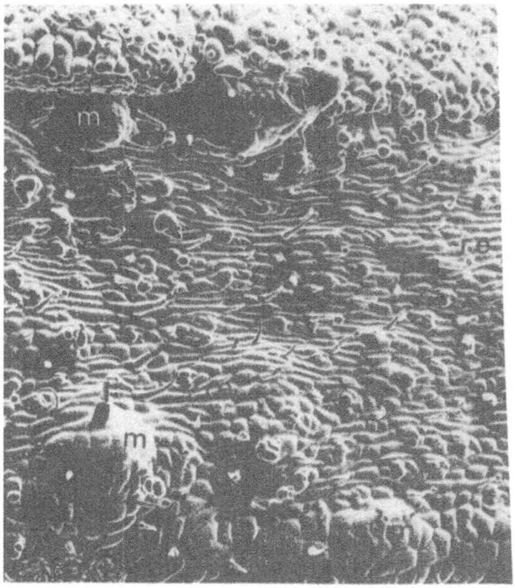

Figure 1. Meristems (m) neoformed at day 5 directly at the surface of a tobacco Thin Cell Layer explant. x 40 (from Tran Thanh Van and Ha Ngoc).

2), vegetative buds, roots (Fig. 3), embryoids and callus can all be induced from the same cell layer, the subepidermal layer, consisting of highly differentiated cells.

Cell Wall Enzymes and Components

Control of morphogenesis by changing the nature or the quantity of cytokinins has led to the isolation and identification of genes involved in flower differentiation (Meeks-Wagner *et al.*, 1989; Neale *et al.*, 1990). Gene families coding for i) cell wall hydrolases (basic chitinase, basic β, 1-3 glucanase), ii) cell wall protein (extensin), iii) stress protein (osmotin) and iv) a yet unknown product, have been activated at an early stage of floral differentiation (day 7). The involvement of cell wall hydrolases (as also shown in carrot somatic embryogenesis) and of their products (cell wall component fragments or oligosaccharides) has been clearly demonstrated on tobacco Thin Cell Layers (Tran Thanh Van *et al.*, 1985; Tran Thanh Van and Mutaftschiev, 1990).

Isoperoxidases

Besides different concentrations of carbohydrates supplied, different ratios of auxin/cytokinin can control differentiation of not only vegetative buds and roots but also flowers and callus. Therefore we have analysed in the soluble fraction of peroxidases, the isoperoxidase pattern and measured the total peroxidase activity in the TCL explants in the donor plant, along the stem gradient from the basis to the apex at the vegetative stage and at the floral stage and

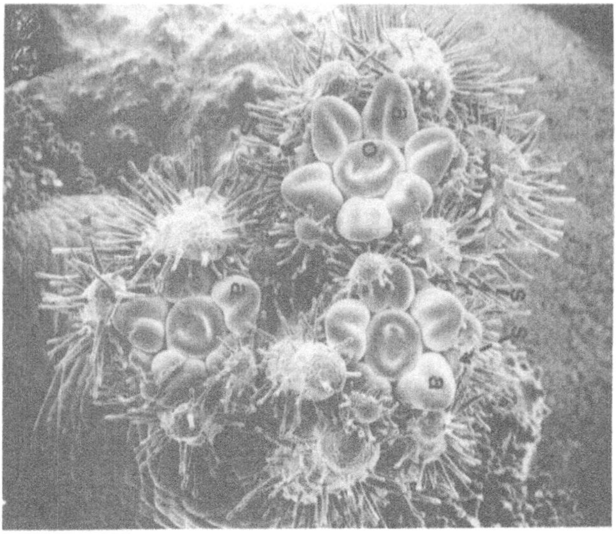

Figure 2. Flowers neoformed at day 10 directly at the surface of a tobacco Thin Cell Layer explant (a: anther, s: sepal, o: ovary) (from Tran Thanh Van and HaNgoc).

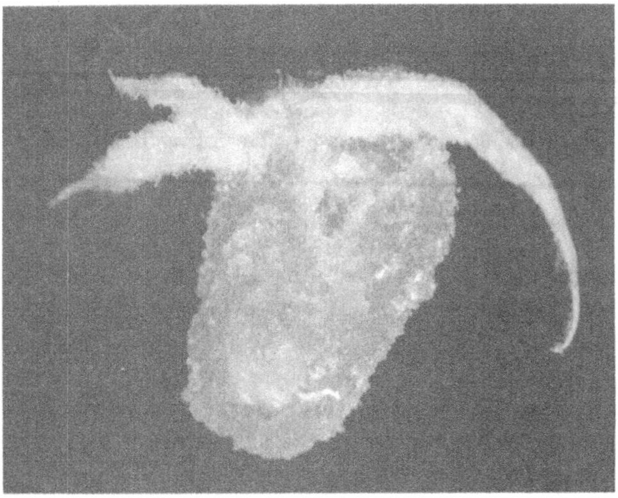

Figure 3. Roots neoformed at day 14 at the surface of a tobacco Thin Cell Layer.

also during the different morphogenic programmes through the course of the *in vitro* culture (Thorpe *et al.*, 1978). The peroxidase activity decreased acropetally in both vegetative and floral stage plants and from the vegetative stage to the floral stage. The number of isoenzyme bands decreased at the floral stage.

Different morphogenic programs were compared for their isoperoxidase activity and isoenzyme patterns: roots (Fig. 4), flowers (Fig. 5), vegetative buds (Fig. 6) and callus (Fig. 7). The root program exhibited the strongest peroxidase activity followed by the flower

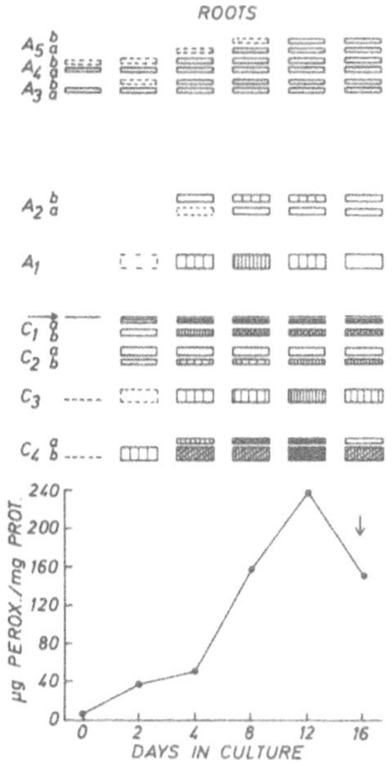

Figure 4. Peroxidase zymogram and changes in total specific peroxidase activity during root formation from tobacco Thin Cell Layers cultured for 16 days. The arrow indicates the time of visible appearance of roots. A: anodic band; C: cathodic band (from Thorpe, Tran Thanh Van, and Gaspar).

program. These activities were approximately 3-and 2-fold the activity of callus and vegetative bud programs. However, there was a decrease in peroxidase activity just after the initiation of roots and of vegetative buds (both at day 12), and at day 8 of the callus program. In contrast, the activity increased with time during flower initiation and development.

As for the isoenzyme pattern, the profile was different for each morphogenic program. The root program exhibited the highest number of isoenzymes. Some of these isoenzymes might be involved in auxin-catabolism and others in the lignification processes.

Up to day 4, the activity was similar for all programs. However, root and callus programs showed marked differences in their isoenzyme pattern. These changes in activity and enzymic isoforms were correlated with specific cytological changes: a first mitotic peak in the subepidermal layer was observed at day 2; a second mitotic peak, at day 6, was followed by tracheid differentiation (Dien and Tran Thanh Van, 1974; Tran Thanh Van and Dien, 1974). It was shown that lignification processes required specific peroxidases (Catesson, 1986).

Other authors analysing both soluble and bound fractions of peroxidase in the tobacco TCL system have shown that there is a correlation between peroxidase isoenzymes, histological changes and the differentiation of flowers, vegetative buds and callus (Kay and Basile, 1987). These authors have identified 3 to 6 isoperoxidases correlated with cell division and/or the differentiation of tracheary elements and 7 for callus formation and stamen development. In *Mercurialis annua*, a stamen specific isoperoxidase was detected by histoimmunolocalisation (Kahlem, 1976).

Table 1. Effects of exogenous putrescine addition on morphogenesis of treated tobacco TCL explants by DFMA, DFMO or MES (after Tiburcio, Gendy, and Tran Thanh Van, 1989).

Medium	Day of tranfer in root medium + 1mM Put	No. organs / explant	Mean % of explant forming				
			Roots	Vegetative buds	Flowers	Floral shoots	Callus
Root		8 ± 1	100	0	0	0	0
Root + MES (50 mM)			0	23	65	55	100
Root + DFMA (0.5 mM)	4	5 ± 1	100	0	0	0	0
	10	3 ± 1	100	0	0	0	0
Root + DFMO (0.5 mM)	4	3 ± 1	0	20 ± 1	0	0	100
	10	1	0	15 ± 1	0	0	100
Root + MES (50 mM)	4	4 ± 1	0	22 ± 2	0	0	100
	10	3 ± 1	0	23 ± 3	0	0	100

Put: Putrescine
DFMA: Difluoromethylarginine
DFMO: Difluoromethylornithine
MES: 2-(N-morpholino) ethane sulphonic acid

Table 2. Effect of exogenous MES, DFMA and DFMO on polyamines titer in tobacco TCL explants (after Tiburcio, Gendy, and Tran Thanh Van , 1989).

Medium	Polyamine titre (nmol g^{-1})			
	Putrescine	Spermidine	Spermine	Putrescine/ (Spermidine+ Spermine)
Root	3119 ± 210	819±130	87±16	3.4
Root+MES (50mM)	734±62b	942±74	158±21	0.67
Root+DFMA (0.5mM)	803±16b	1498±116	230±37	0.46
Root+ DFMO (0.5mM)	2275±139	1653±110a	215±5a	1.2

Table 3. Polyamine levels in flower- (F) and vegetative buds-(VB) forming TCL of tobacco.

Age of culture (day)	Titers (nmol/mg proteine) Putrescine			Spermidine		
	VB	F	F/VB	VB	F	F/VB
0	88 ± 9			83 ± 12		
9	249 ± 29	167 ± 15	0.63	201 ± 24	241 ± 1	1.19
15	331 ± 42	227 ± 24	0.68	189 ± 22	255 ± 17	1.35
20	769 ± 59	834 ± 62	1.09	217 ± 21	436 ± 73	1.96
28	879 ± 84	1128 ± 69	1.28	188 ± 20	865 ± 20	4.54

Table 4. Effect of cyclohexylamine on spermidine titer and floral differentiation in tobacco TCL explants (after Kaur-Sauhney, Tiburcio, and Galston (1988).

Cyclohexylamine(mM)	Spermidine titer (nmol /g Fresh Weight)	No. Organs/12 Explants Flowers	Vegetative buds
0	841 ± 93	21	0
2	530 ± 174	12	4
10	388 ± 71	5	8
20	216 ± 30	2	31

initial pH of 5.83. In these experiments, the medium pH decreased from 5.83 to 5.04. When this medium was buffered by adding MES (50 mM), roots were not formed; instead vegetative buds, flowers, floral shoots as well as callus were obtained (Table 1).

We have shown (in collaboration with Tiburcio) that root differentiation is correlated with a peak of putrescine synthesis at day 7. Inhibition of putrescine synthesis was shown to be more affected by DFMA (0.5 mM) than by DFMO (0.5 mM) suggesting that putrescine was synthezised via an arginine decarboxylase pathway (Table 2). The ratio putrescine / (spermidine and spermine) levels decreased from 3.4 to 0.46 and in parallel, root differentiation was inhibited (Table 1). Addition of putrescine or of spermidine (1 mM) at day 4 and day 10 of culture was followed by root differentiation only when putrescine (and not spermidine) was supplied. However the sole addition of putrescine to a medium which was buffered was not sufficient to induce root formation (Table 1). A decrease of pH during the time course of culture and especially at day 7 was also required (Cousson *et al.*, 1989 ;

Gendy, 1991). These data suggest that polyamines, in this case putrescine, may act in a specific manner, independently of any modulating effect upon pH via their role as organic carriers of positive charges. These results obtained by using the TCL system (which is only composed of epidermal, subepidermal and cortical cells which normally do not initiate roots) support the hypothesis that putrescine metabolism may be involved in root differentiation (Tiburcio *et al.*, 1989).

Spermidine as Marker for Flower Differentiation

Results obtained by Kaur - Sawhney *et al.* (1988) showed that endogenous spermidine level was higher in flower-forming-TCL compared to vegetative bud-forming-TCL. The ratio between spermidine level of flower-TCL and vegetative bud-TCL was 4.54 (Table 3). As for putrescine, this ratio was only 1.28. Similarly to putrescine for root differentiation mentioned above, inhibition of spermidine synthesis by increasing concentrations of cyclohexylamine (20 mM), a specific inhibitor of spermidine synthase, was followed by the inhibition of flower differentiation and resulted in the differentiation of vegetative buds (Table 4). Exogenous supplies of increasing concentrations of these different polyamines showed that only spermidine can reverse this inhibition, although only partially (Table 4). It seems that spermidine and putrescine synthesis respectively parallel flower and root differentiation and that they may act specifically. Other authors (Altamura *et al.*, 1991) have also shown that *de novo* root formation on tobacco TCL is affected by the inhibition of polyamines biosynthesis.

Canellakis *et al.* (1985) showed that spermidine was taken up and bound to a specific protein of 30 kD in fertilized eggs of sea urchin during early stages of embryogenesis. Apelbaum *et al.* (1988) have also shown that in tobacco TCL induced to form flowers, spermidine was bound to a protein of approximately 18 kD. Based on these results, the authors suggested possible postranslational modifications of proteins by polyamines.

In their review (this volume), Bagni *et al.* have stipulated that polyamines are involved in cell division and not in morphogenesis. Furthermore, according to these authors, added polyamines are ineffective in modifying morphogenic processes. Our results (Gendy 1991; Tiburcio *et al.*, 1989) demonstrated clearly that the peak of biosynthesis of putrescine at day 7 was correlated with a specific type of differentiation (root differentiation) and not only to cell division, for cell division occurs also in the control TCL which exhibits differentiation of flowers, vegetative buds as well as callus.

Although an involvement of polyamines biosynthesis in tobacco TCL organogenesis has been confirmed by several authors including Bagni (Altamura *et al.*, 1991), the difference between the results presented in this volume by Bagni *et al.*, on one hand and our results (as well as other authors' results, Altamura *et al.*, 1991; Appelbaum *et al.*, 1988; Canellakis *et al.*, 1985; Kaur-Sawhney *et al.*, 1988) on the other hand (all dealing with tobacco TCL), could be partly explained by the use of different starting material; that is, stem internode TCL and pedicel TCL to generate respectively vegetative buds and flowers whilst only floral branches TCL will generate all of the morphogenic programmes in our study. It was demonstrated previously by Tiburcio *et al.* (1988) that the endogenous level of polyamines varied with the type of organs selected and the physiological stage of the donor plant (Martin-Tanguy *et al.*, 1987) at which the TCLs were excised. We are aware of the fact that growth substances change the level of endogenous polyamines content. We have made the choice of starting with identical polyamine levels and of studying variations in these levels induced by growth substances, being selected to induce specific morphogenic patterns with appropriate TCL controls for i) wound effects , ii) absence of morphogenesis and iii) cell division.

Plant Cell Wall Oligosaccharides

It was clearly demonstrated for first time by our group (in collaboration with Albersheim's group) that pectic components released from plant cell walls by purified endopolygalacturonases (EPG fragments) can control organ differentiation, cell division and cell enlargement in the tobacco TCL system (Tran Thanh Van *et al.*, 1985; Tran Thanh Van and Mutaftschiev, 1990).

In a next step, we investigated the influence of oligosaccharides in relation to pH variation of the culture medium. In order to study the mechanism of action of oligosaccharides, further purification of the biologically active oligosaccharides was required. This purification could not be done without a sensitive experimental model. The tobacco TCL has allowed V. Marfa (from our group) to undertake this purification in Albersheim's laboratory. They have shown that oligogalacturonides purified from sycamore cell wall (DP 12-14) are active on flower differentiation and that the oligogalacturonides of DP< 10 exhibited no activity or only low activity. Similar results were obtained with oligogalacturonides obtained from other sources such as those of tobacco (Marfa *et al.*, 1991) and on oligosaccharides from algae cell walls (unpublished results, in collaboration with J.C. Yvin and F. Levasseur).

A sulfated tetraglucosamine was identified as a signal molecule in the triggering of cortical cell division leading to root nodule differentiation in the rhizobium / leguminous symbiotic interaction. This molecule is active at concentrations as low as 10^{-11}M (Lerouge *et al.*, 1990).

In collaboration with P. Sinai, we have tested the biological activity of several synthetic oligosaccharides. The optimum concentration of a synthetic tetrasaccharide to be active in inducing in the wheat coleoptile an elongation equivalent to that induced by 10^{-6} M IAA was as low as 10^{-11}M. The same oligosaccharide was also shown to be active at 10^{-8} M in inducing changes in morphogenesis of tobacco TCL and especially for inducing cell division. The biological effect of these synthetic oligosaccharides was thus established as being respectively 100,000 or 100 times more active than IAA (for coleoptile elongation) or IAA and kinetin (for tobacco TCL organogenesis) (Gendy, 1991; Tran Thanh Van, 1991).

CONCLUDING REMARKS

Although a wide range of factors, molecules and compounds can trigger cell division, cell expansion and morphogenesis, they may act through a much narrow range of bio-

Neale A, Walhleithner J, Lund M, Bonnet, Kelly A, Meeks-Wagner D, Peacock W, Dennis E (1990) Chitinase, β - 1, 3 - Glucanase, osmotin, and extensin are expressed in tobacco explants during flower formation. *Plant Cell* 2: 673-684.

Nomura K, Komamine A (1985) Identification and isolation of single cells that produce somatic embryos at a high frequency in a carrot suspension culture. *Plant Physiol* 79: 988-991.

Polissier B, Bouchefra O, Popin R, Freyssinet G (1990) Production of isolated somatic embryos from sunflower thin cell layers. *Plant Cell Reports* 9: 47-50.

Raskin L (1992) Role of salicylic acid in plants. *Annu Rev Plant Physiol Plant Mol Biol* 43: 439-464.

Rosen SD, Kafka J, Simpson DL, Barondes SH (1973) Developmentally-regulated, carbohydrate-binding protein in *Dictyostelium discoideum. Proc Natl Acad Sci USA* 70: 2554-2557.

Smith TA (1985) Polyamines. *Annu Rev Plant Physiol* 36: 117-143.

Steward FC, Mapes MO, Smith J (1958) Growth and organised development of cultured cells I. Growth and division of freely suspended cells. *Am J Bot* 45: 693-703.

Tabor CW, Tabor H (1985) Polyamines in microorganisms. *Microbiol Rev* 49: 81-99.

Thorpe T, Tran Thanh Van K, Gaspar T (1978) Isoperoxidases in epidermal layers of tobacco and changes during organ formation *in vitro. Physiol Plant* 44: 388-394.

Tiburcio AF, Kaur - Sawhney R, Galston A (1988) Polyamine biosynthesis during vegetative and floral bud differentiation in thin layer tobacco tissue cultures. *Plant Cell Physiol* 29: 1241-1249.

Tiburcio AF, Gendy A, Tran Thanh Van K (1989) Morphogenesis in tobacco subepidermal cells: putrescine as marker of root differentiation. *Plant Cell Org Tiss Cult* 19: 43-54.

Tran Thanh Van M (1973a) Direct flower neoformation from superficial tissues of small explant of *Nicotiana tabacum* L. *Planta* 115: 87-92.

Tran Thanh Van K (1973b) *In vitro* and *de novo* flower, bud, root and callus differentiation from excised epidermal tissue. *Nature* 246: 44-45.

Tran Thanh Van K, Dien NT (1975) Etude au niveau cellulaire de la différenciation *in vitro* et *de novo* de bourgeons végétatifs, de racines, ou de cal a partir de couches minces de cellules de type épidermique de *Nicotiana tabacum* Wisc. 38. *Can J Bot* 53: 553-559.

Tran Thanh Van K (1980) Control of morphogenesis by inherent and exogenously applied factors in thin cell layers. *Int Rev Cytol* 11 A: 175-194.

Tran Thanh Van K, Toubart P, Cousson A, Darvill AG, Gollin DJ, Chelf P, Albersheim P (1985) Manipulation of the morphogenetic pathways of tobacco explants by oligosaccharins. *Nature* 314: 615-617.

Tran Thanh Van K, Mutaftschiev S (1990) Signals influencing cell elongation, cell enlargement, cell division and morphogenesis. *In*: Progress in Plant Cellular and Molecular Biology, Nijkam HJJ, Van Der Plas LHW, Van Aartij J (eds), Kluwer Academic.

Tran Thanh Van K, Richard L, Gendy C (1990) An experimental model for the analysis of plant / cell differentiation: Thin cell layer concept, strategy, methods, records and potential. *In*: Plant Aging, Basic and Applied Approaches, Rodriguez R, Tames RS, Durzan D (eds), Plenum Press, New York.

Tran Thanh Van K (1991) Molecular aspects of flowering. *In*: Genetics and Breeding of Ornamental Plants, Harding J, Singh F, Mol JNM (eds), Plenum Academic Press.

Vasil V, Vasil AK (1982) Characterization of an embryogenic cell suspension culture derived from cultured inflorescences of *Pennisetum americanum* (pearl millet, *Gramineae*). *Amer J Bot* 69: 1441-1449.

WHAT IS THE SIGNIFICANCE OF THE RELATIVELY LARGE TUBULIN MULTIGENE FAMILIES FOR PLANT MORPHOGENESIS?

Donald E. Fosket, Hiroshi Tonoike, In-Seob Han and **Adán Colón**

Dept. of Developmental and Cell Biology
University of California
Irvine, California 92717, USA

INTRODUCTION

The fibrous subcellular structures known as microtubules (MTs) are a major component of the cytoskeleton. They are found in most cell types of virtually all eukaryotic organisms where they play vital roles in cell function and in cellular morphogenesis. They are necessary for such typical eukaryotic cell behavior as mitosis and meiosis, where they provide the mechanism for chromosome movement, and they are responsible for intracellular vesicle transport and for some types of cell motility. Chromosome movement and MT-dependent vesicle transport are brought about by mechanoenzymes, such as dynein or kinesin, which attach to vesicles or kinetochores and move along the surface of the MT, pulling these structures with them. There appears to be a great deal of similarity if not identity in the components of these motility systems throughout the eukaryotes. The molecular composition and function of the mitotic spindle, for example, is similar in angiosperm and other groups of eukaryotes. However, other higher plant MT arrays appear to have unique functions that have no precise counterpart in other groups of organisms. This is particularly true of the cortical MTs that are involved in orienting the deposition of cellulose microfibrils and of the PPB that appears to control the plane of cytokinesis. Although the phragmoplast is analogous to the midbody of vertebrates in that both participate in cytokinesis, it is substantially different from the midbody in function. The phragmoplast MTs participate in constructing the new cell wall that partitions the dividing cell.

Each of these cellular functions is believed to occur through the interaction of microtubule-associated proteins (MAPs) with these MTs. The interaction of MAPs with MTs is responsible for their functional properties. Unfortunately, at the present time very few angiosperm MAPs have been characterized biochemically, although they are visualized readily by electron microscopy. The structure of MTs is highly conserved during the

Morphogenesis in Plants, Edited by K.A. Roubelakis-Angelakis
and K. Tran Thanh Van, Plenum Press, New York, 1993

evolution of the eukaryotes, as are the tubulin proteins from which MTs are constructed. Tubulins are encoded by multigene families in those angiosperms that have been examined and most cell types contain the products of several α– and β–tubulin genes. Although some tubulin genes are expressed in many tissues and organs, others are more limited in their expression, or are more tightly regulated, so that these tubulin isotypes are found only in a few cell types or in a specific organ at a particular stage in its development. Here we review the evidence that the interaction of MAPs with microtubules is controlled by the variable carboxyl-termini of both α– and β–tubulins. This region of sequence variability, which defines the different tubulin isotypes, is part of a larger functional region of the tubulin protein that has been termed the regulatory domain and has been shown to contain the binding sites for calcium ions and MAPs, both of which strongly affect MT assembly. Certain MAPs may bind only to MTs containing a specific tubulin isotype. As a result, the appearance of a specific tubulin isotype in a given cell could enable the MTs of that cell to carry out some function they would be incapable of in its absence.

THE STRUCTURE OF MICROTUBULES IS VERY SIMILAR IN NEARLY ALL EUKARYOTES

Microtubules are proteinaceous hollow rods with a diameter of approximately 24 nm and a variable length. Their structure is remarkably similar in a wide spectrum of eukaryotic organisms from diverse phyla. They are formed by the regulated self-assembly of the protein, tubulin. The structural protein of MTs is a heterodimer with a mass near 100,000. It is composed of an α- and β-tubulin, each of which have a mass of approximately 50,000. The α- and β-tubulin monomers are held together in the heterodimer by non-covalent, hydrophobic interactions. The dissociation constant for the dimer is very low. As a result, cellular tubulin is almost all in the form of soluble dimers and MTs (Sackett and Lippoldt, 1991). X-ray diffraction studies at 18 Å resolution suggests that each tubulin monomer consists of two compact domains and a projection domain which protrudes from one end of the molecule (Beese *et al.*, 1987). Tubulin self assembly occurs as a result of two types of self recognition and hydrophobic bonding (Fig. 1). First, tubulin dimers bind head-to-tail to form linear protofilaments. Secondly, the protofilaments interact laterally to form the MT wall. The dimers of adjacent protofilaments are offset to form a three start, left-handed α–helix with a 10° helical pitch (Figure 1C) (Mandelkow *et al.*, 1986). Although most MTs contain 13 protofilaments and have a diameter of 24 nm, certain specialized cells may contain as few as 11 or as many as 15 protofilaments. *In vitro* assembly of tubulin under different solution conditions also can result in the formation of MTs with a variable number of protofilaments or the formation of other polymeric structures (Karecla *et al.*, 1989; Scheele *et al.*, 1982). The high degree of conservation of MT structure is strongly suggested by the fact that tubulins from divergent organisms will co-assemble to form MTs. For example, when mammalian brain tubulin is injected into *Tradescantia* stamen hairs, it is incorporated into all the MT arrays of these angiosperm cells (Zhang *et al.*, 1990). Additionally, *Physarum* tubulin is incorporated at random into the fibroblast MTs (Prescott *et al.*, 1989). However, the incorporation of the *Physarum* tubulin into fibroblast MTs was shown to alter their sensitivity to colchicine.

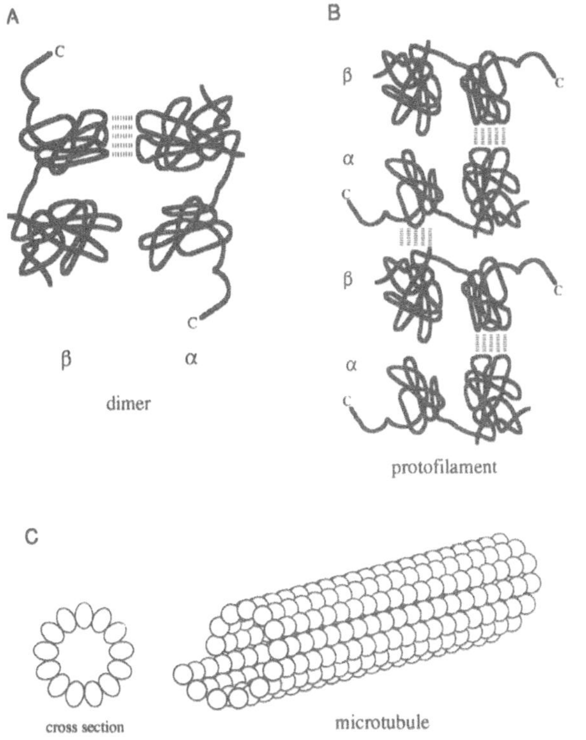

Figure 1. The structure of tubulin monomers, dimers, protofilaments and microtubules. The pattern of tertiary folding of tubulins is not known, but both α- and β-tubulins exhibit the two compact domains and a more linear carboxyl-terminal domain, represented in (A). Dimer formation occurs through hydrophobic bonds between an α- and β-tubulin monomer. Hydrophobic bonding also is responsible for dimer aggregation in the formation of protofilaments (B) and for dimer interactions leading to microtubule assembly (C).

TUBULIN AMINO ACID SEQUENCES ARE HIGHLY CONSERVED IN EVOLUTION

The three different tubulin proteins that have been identified thus far have been designated α-, β- and γ-tubulin. Each of these tubulins shares 35 - 40% amino acid sequence identity with the other two (Burns, 1991). The tubulin heterodimer which self-assembles to form MTs is composed of α– and β–tubulin monomers only. Gamma–tubulin is not a structural protein of MTs, but a relatively rare protein that is found in the centrosome of higher eukaryotes and the spindle pole body of fungi (Oakley *et al.*, 1990; Stearns *et al.*, 1991; Zheng *et al.*, 1991). Gamma-tubulin may function in the initiation of MT assembly. The genes encoding α– β– and γ–tubulin apparently arose early in the evolution of the eukaryotes as a result of the amplification of an ancestral tubulin gene. This initial amplification was followed by the rapid divergence of the copies to establish the progenitor genes from which all modern tubulins are derived, Since this initial divergence, tubulins have undergone very slow and limited evolutionary divergence. Present day organisms from yeast

Table 1. Amino acid sequences of the carboxyl-terminus of some of the β-tubulins expressed in different soybean organs. Sequences were derived from clones obtained by PCR of cDNAs synthesized from mRNA extracted from different soybean organs. Each sequence is preceded by the clone designation from which it was derived and the sequence is given using the single letter code.

Sequence of the carboxyl terminus of previously cloned soybean β-tubulin genes
SB-1 AEYQQYQDAT AVDDHEDEDE DEAMAA*
SB-2 SEYQQYQDAT ADEDEYEEEE EEEEFAQHDM*
SB-3 SEYQQYQDAT AEDDGEYEDE EDDDVEADHM*
β-tubulins expressed in enlarging leaves
EL-6 AEYQQYQDAT ADEEEYEEEE EEEEVGA*
El-11 SEYQQYQDAT AEDDGEYEDE EDEQM*
EL-40 SEYQQYQDAT AEDDGEYEDE EDDDVEADHM*
β-tubulins expressed in mature leaves
M-14 SEYQQYQDAT AEDEGEYEDE REM*
M-28 SEYQQYQDAT AEDDGEYEDE EEEDGEADDH M*
β-tubulins expressed in etiolated internodes
I-10 SEYQQYQDAT AEDDGEYEDE EDNQSEG*
I-18 HEYQQYQDAT AEDEGEYEDE EEEDGEADDH M*
I-33 SEYQQYQDAT AEDDGEYEDE EDEDGDADDH*

variable region (Fosket and Morejohn, 1992; Sullivan, 1988). There is a high degree of conservation of specific β–tubulin isotypes among different vertebrate species. For example, five of the six β–tubulin isotypes are 97-99% identical in mouse and chicken (Sullivan, 1988; Wang *et al.*, 1986). In fact, more sequence variability is seen among the isotypes of a given vertebrate than there is when the same isotype is compared from mouse and chicken. The strong conservation of β–tubulin isotypes among vertebrates suggests that individual isotypes may be specialized functionally to perform some unique cellular task, for which the other isotypes cannot substitute. However, the α–tubulin isotypes do not exhibit a similar interspecies conservation of carboxyl-terminal sequence. Furthermore, at the present time there is no evidence for the conservation of any isotypic sequence among the angiosperm tubulins, although we may have too few sequences from different angiosperm species to make this comparison at this time.

ANGIOSPERM TUBULINS CONTAIN SOME UNIQUE AMINO ACID SUBSTITUTIONS

Although there is an overall similarity in the structure and sequence of vertebrate and angiosperm tubulins, there are some consistent differences as well. The most striking of these are several positions outside the hypervariable carboxyl-terminus in which plant tubulins exhibit additional hypervariability. These are positions in which there is either no consensus or a very weak consensus in the plant tubulins, where there is complete conservation or a strong consensus for a particular amino acid in the vertebrate tubulins (Fosket and Morejohn, 1992). Two of these hypervariable positions occur within the amino terminal region of diversity and two of the *Arabidopsis* β–tubulins have an additional residue inserted in this hypervariable region (Snustad *et al.*, 1992). However, other hypervariable positions occur in regions of the protein that show a very high degree of sequence conservation in the vertebrate tubulins. Additionally, there are positions in which plant α– or β–tubulins all have one amino acid while the vertebrates all have a different amino acid (Fosket and Morejohn, 1992). Plant α–tubulins have more than twice as many of these plant-specific substitutions than the β–tubulin, and more of these are non-conservative. These differences between the angiosperm and vertebrate tubulins could lead to important differences in the tertiary structure of these proteins. Since both angiosperm and vertebrate tubulins form MTs with a very similar structure, these differences are less likely to affect the parts of the protein that participate in MT assembly than in the interaction of MTs with MAPs or other molecules. The fact that these differences have been selected and preserved during the evolution of plants and animals from a common ancestor suggests that they may optimize the function of the plant or animal tubulins in these organisms.

ANGIOSPERM TUBULINS DIFFER FROM THOSE OF VERTEBRATES IN SOME OF THEIR PHYSIOLOGICAL AND PHARMACOLOGICAL PROPERTIES

The MT arrays of plants differ from those of vertebrates in their sensitivity to cold and to certain pharmacological agents. Vertebrate brain MTs are highly susceptible to depolymerization by cold. Indeed, cyclic warm-induced assembly, followed by cold induced depolymerization, has been used as a method for the isolation of MT proteins (Borisy *et al.*, 1975), and cellular MTs can be depolymerized by cold. The MTs of many plants tend to be more resistant to cold, at least *in vivo*. Cortical MT arrays are disassembled in the cold in cells of some tropical plants such as *Hemanthus*, but MTs of temperate plants such as rye are rather resistant to cold depolymerization and many cytoplasmic MTs survive at temperatures near freezing (Kerr and Carter, 1990). This may not be an inherent property of the tubulin, however, but rather a reflection of stability imparted on the microtubule by its interaction with MAPs.

Angiosperm and vertebrate tubulins also differ in their response to pharmacological agents and to certain herbicides. Some classes of herbicides bind to plant tubulins with high affinity, and vertebrate tubulins completely lack these high affinity binding sites. Additionally, plant tubulins are much less sensitive to the classic anti-MT agent, colchicine, than vertebrate tubulins. Vertebrate tubulins have three different, well characterized, high-affinity binding sites for anti-MT drugs; a colchicine/podophyllotoxin site, a vinblastine/maytansine site, and a taxol binding site. At the present time we do not know if herbicides bind to one of these already characterized ligand binding sites or if, during the evolution of angiosperm tubulins, they acquired a new ligand binding site with an affinity for certain herbicides. Since the kinetics of colchicine binding to angiosperm tubulin suggest that the angiosperm colchicine/podophyllotoxin binding site has diverged from that of the vertebrate tubulins, it is possible that this site on angiosperm tubulin was modified during evolution so that it acquired an affinity for herbicides. This would be completely fortuitous, since it is unlikely that angiosperms were exposed to these chemically synthesized molecules until very recent times.

The anti-MT drugs, colchicine, taxol, podophyllotoxin and vinblastine, are all alkaloids and plant products. Cellular compartmentation probably prevents the interaction of these agents and the MTs of the cells producing them. Taxol is a complex taxane alkaloid that is produced by plants in the genus *Taxus*. It was identified as an anti-leukemic agent (Wani *et al.*, 1971), and it is unique in that it blocks MT-dependent processes by binding to MTs and preventing tubulin disassembly. Treatment of cells with micromolar levels of taxol leads to the stabilization of existing MTs and the formation of additional MTs in both higher plants and other organisms (Bajer *et al.*, 1982; Baum *et al.*, 1981; De Brabander *et al.*, 1981).

Taxol lowers the critical concentration required for assembly of both plant and vertebrate tubulins and it permits tubulin assembly *in vitro* in the absence of MAPs and GTP (Morejohn and Fosket, 1984a; Schiff *et al.*, 1979). Taxol binds to polymerized tubulin in MTs and not to the unpolymerized tubulin dimer (Parness and Horwitz, 1981). Apparently the taxol binding site is created by an assembly-dependent change in tubulin configuration. The binding of taxol to MTs renders them resistant to depolymerization by anti-MT drugs and other agents, such as calcium or cold (Kumar, 1981; Parness and Horwitz, 1981; Schiff and Horowitz, 1981; Schiff *et al.*, 1979). The overall similarity of the response of plant and animal MTs to taxol suggests that taxol-binding site on angiosperm MTs is highly conserved (Morejohn and Fosket, 1984a).

The alkaloid colchicine is produced by members of the *Liliaceae*, including all species of *Colchicum*, and several species of *Bulbocodium* (Eigsti and Dustin, 1955). Colchicine is an anti-MT agent for both angiosperms and vertebrates, but the response of plant cells to colchicine differs markedly from that of vertebrate cells. The most striking difference is the fact that prolonged treatment with colchicine usually kills animal cells, even in the micromolar doses used to block mitosis (Deysson, 1975). Plant cells, on the other hand, usually are not killed by colchicine treatment, even at high concentrations. Colchicine depolymerizes MTs of the mitotic spindle, but dividing cells survive for days in a state of colchicine-induced mitotic arrest, and still resume growth and division as polyploids after the colchicine is removed (Blakeslee and Avery, 1937; Levan, 1938). Plants and vertebrates also differ in their sensitivity to colchicine. Most vertebrate cell MTs are depolymerized rapidly by sub-micromolar colchicine concentrations, while millimolar levels of colchicine are required to achieve the same effect in plant cells (Deysson, 1975). There is also considerable species variability in the response of plant cells to colchicine and the MTs of some species, particularly colchicine-producing plants, are completely resistant to colchicine (Levan and Steinegger, 1947).

The vertebrate tubulin dimer has one high-affinity colchicine binding site (Hains *et al.*, 1978). Colchicine binds to angiosperm tubulins, but for most species tested, the affinity of this binding is approximately 1,000 fold less than that of vertebrate tubulin (Morejohn *et al.*, 1987b). The affinities of vertebrate and angiosperm tubulins for colchicine are very different and this difference is the most likely explanation for the differing sensitivities of MTs in these two groups of organisms to colchicine-induced depolymerization. This may account as well for some of the differences in colchicine sensitivity among different angiosperm species since the tubulins of different plant species also differ in their affinity for colchicine (Morejohn *et al.*, 1984). Carrot tubulin exhibits a moderate affinity for this alkaloid, but it is still over ten fold lower than the affinity of vertebrate tubulins (Kato *et al.*, 1985).

The dinitroaniline herbicides, which include trifluralin and oryzalin, and phosphoric amide herbicides, including cremart and amiprophos methyl (APM), are used to control annual grasses and broadleaf weeds in various crops (Aya *et al.*, 1975). Dicots tend to be more resistant to the dinitroaniline herbicides than monocots for reasons that have not been determined, but these compounds have few effects on vertebrates at concentrations that have profound effects on plants. Many of the effects of the dinitroaniline and phosphoric amide herbicides on plants can be explained on the assumption that these compounds are anti-MT agents and that MTs are their primary cellular target (Morejohn and Fosket, 1992). For example, the effects of these herbicides include the arrest of mitosis in metaphase, a classic effect of anti-MT drugs, and a loss of cytoplasmic microtubules (Jackson and Stetler, 1973; Kiermayer and Fedtke, 1977; Verhoeven *et al.*, 1990). Treatment of plant cells with low concentrations of oryzalin leads to the rapid loss of MTs from most arrays, and this effect is reversible when the herbicide is washed out (Morejohn *et al.*, 1987a). The rapidity of this response and sensitivity of plant MTs to these agents is illustrated by the fact that treatments as short as 2 minutes with 1 micromolar oryzalin result in the loss of nearly all of the radial array MTs from *Hemanthus* endosperm cells, with only short MT stubs remaining attached to the nuclear envelope (Bajer and Mole-Bajer, 1986). The same treatment also brought about the rapid loss of phragmoplast and kinetochore MTs from dividing *Hemanthus* cells. Prolonged dinitroaniline herbicide treatment also results in gross morphological abnormalities in the tips of growing roots, as well as the inhibition of primary and lateral root growth. These abnormalities include the swelling of the root in the region of cell elongation as a result of the radial expansion of the enlarging cells, a phenomenon similar to the "*c*-tumors" that are formed in roots after prolonged colchicine treatment (Cleary and Hardham, 1990; Upadhyaya and Nooden, 1978). These effects also illustrate the important role of MTs in determining plant cell morphology, as presented below.

Oryzalin has been shown to inhibit angiosperm tubulin assembly, but not bovine brain tubulin assembly, *in vitro*, and to bind to rose tubulin with moderately high affinity (Morejohn *et al.*, 1987a). It has also been shown to bind to *Chlamydomonas* tubulin and to lack affinity for vertebrate tubulins (Strachan and Hess, 1983). APM also inhibits the assembly of angiosperm tubulin *in vitro*, but it is without effect on the bovine brain tubulin assembly (Morejohn and Fosket, 1984b). Propyzamide is another herbicide structurally unrelated to the phosphoric amide or dinitroaniline herbicides, which has been shown to block angiosperm, but not vertebrate tubulin assembly *in vitro*, and to act as an anti-MT agent in plant cells *in vivo* (Akashi *et al.*, 1988). Thus, tubulin appears to be a primary, if not the only, target for the cellular effects of these herbicides. Additionally, these results point to significant differences between vertebrate and angiosperm tubulins in their ligand binding properties.

CELLULAR MICROTUBULE ARRAYS ARE DYNAMIC

MTs in mammalian cells turn over rapidly, with half lives on the order of minutes (Cassimeris *et al.*, 1988). Although the half life of MTs in plant cells has not been determined, the fact that the PPB, phragmoplast and mitotic spindle arrays are present in cells for only few hours suggests that the half lives of the individual MTs of these arrays may be very short. The MTs of the cortical array may be more stable than those of dividing cells, but even here it is likely that there is fast turnover of individual MTs since labeled tubulin is rapidly incorporated into cortical MTs after it is injected into plant cells (Zhang *et al.*, 1990). MTs generally are in a dynamic equilibrium with a pool of soluble tubulin dimers so that individual MTs grow slowly by the addition of subunits from the soluble pool for a period of time and then suddenly and rapidly depolymerize (Mitchison and Kirschner, 1987).

The tubulin dimer contains all the information necessary to form MTs and tubulin dimers will self-assemble *in vitro* in the absence of any other proteins (Bayley and Manser, 1985). Tubulin self assembly is a two step reaction. It is initiated by a highly exothermic, comparatively slow nucleation reaction which is followed by the addition of subunits to the nucleus to elongate the MT (Jackson and Berkowitz, 1980; Johnson and Borisy, 1977; Voter and Erickson, 1984). Because the nucleation reaction is rate limiting for self assembly, MT formation occurs much more readily when other proteins or structures are present to act as nucleation sites, or MT organizing centers (MTOCs) All cellular MT formation appears to be nucleated and the nucleation step is the point at which MT formation is regulated. The centrosome is the most important nucleator of the MTs of the mitotic spindle in plants and animals and of the interphase array in vertebrate cells. In animal cells the centrosome consists of a centriole or centriole pair and a surrounding cloud of amorphous electron dense material known as pericentriolar granules. Microtubules are nucleated by the pericentriolar material (Gould and Borisy, 1977; Mitchison and Kirschner, 1987). Although angiosperms lack centrioles, the mitotic spindle of plant cells appears to be nucleated by structures that are analogous to pericentriolar granules (Pickett-Heaps, 1969). The dividing cells of bryophytes and vascular cryptogams have a single plastid which acts as the centrosome. MTs are nucleated from the surface of the plastid. Division of the plastid and the migration of the daughter plastids to the poles precedes and accompanies the formation of the mitotic and meiotic spindle. Plastid migration is dependent upon the MTs nucleated from their surface, as is the formation of the spindle itself (Brown and Lemmon, 1990; Busby and Gunning, 1989).

Biochemically, the animal centrosome is complex. In *Drosophila* at least 14 proteins with an affinity for MT have been localized to the centrosome (Kellogg *et al.*, 1989). Gamma-tubulin is an important component of the centrosome in fungi, animals and possibly higher plants (Stearns *et al.*, 1991). The nucleation of MTs by the centrosome appears to be regulated by phosphorylation and dephosphorylation of its component proteins during the cell cycle (Centonze and Borisy, 1990; Vandre *et al.*, 1984).

MTOCs are not well characterized in plant cells, but the strong evolutionary conservation of the mitotic mechanism suggests that many aspects of spindle MT nucleation would be shared by angiosperms and other groups of organisms. Perhaps phosphorylation-dephosphorylation of MTOC determinants also accounts for the fact that tobacco nuclei will nucleate MT assembly *in vitro* if they are isolated from G2 phase cells, while nuclei isolated from G1 phase cells do not (Hasezawa *et al.*, 1991). The human autoimmune serum 5051, which has been used to study the centrosome in mammalian cells (Calarco-Gillam *et al.*, 1983), also stains granules and the surface of bryophyte plastids that nucleate the mitotic spindle MTs in these cells (Busby and Gunning, 1989; Clayton *et al.*, 1985; La Claire and Goddard, 1989). This antiserum also reacts with material at the cytoplasmic face of the nuclear envelope which could be involved in nucleating MTs of the radial array. However, the 5051 antiserum does not react with other structures in interphase cells and thus cannot explain the nucleation of cortical MTs. Either the MTOCs nucleating the cortical MT array are too diffuse to be detected by immunocytochemistry, or they differ from MTOCs nucleating the mitotic spindle. Apparently 5051 antiserum is directed to several epitopes (Harper *et al.*, 1989), so determining the nature of the reactive components responsible for MT nucleating activity will not be straightforward.

MICROTUBULES ARE IMPORTANT DETERMINANTS OF PLANT DEVELOPMENT

Microtubules play at least five different roles in plant morphogenesis: 1) They are involved in the establishment and maintenance of cell polarity. 2) They may control the distribution of plastids and other organelles within the cell. 3) They could bring about the localization of cytoplasmic determinants of cell differentiation during division. 4) They determine the plane in which cytokinesis occurs. 5) Microtubules are the most important determinants of cell shape, usually through their role in orienting the deposition of cellulose microfibrils.

MTs Participate in the Establishment and Maintenance of Cell Polarity

Many plant cells are highly polarized and the establishment of this polarity occurs early in the differentiation of cells from meristem derivatives. Unfortunately, relatively little is known about the mechanisms responsible for the establishment of angiosperm cell polarity, although the phenomenon of polarity has been studied extensively, particularly in lower plants (Quatrano, 1978; Schnepf, 1986). However, in some cases it has been established that the cytoskeleton plays a key role in the development or maintenance of cell polarity. This is most apparent where polarity changes in response to wounding, or during the establishment of a new growth axis. The succulent *Graptopetalum paraguayense* will form new shoot apical meristems at the base of mature leaves. The cells participating in the formation of these meristems initially are polarized along the axis of the leaf. As the meristem is formed, many of the epidermal cells participating in meristem formation change their polarity so that they are oriented at right angles to that of the longitudinal axis of the leaf. Hardham and her co-workers demonstrated that MTs of the cortical cytoskeleton, as well as the cellulose MFs in the cell wall, become realigned to conform with the new growth axis before the meristem initiates active growth (Hardham *et al.*, 1980). When tissues consisting of highly polarized, mitotically quiescent parenchyma cells are wounded, some of the cells in the vicinity of the wound are stimulated to divide, enlarge and differentiate (Sinnott and Bloch, 1945). Many of the wound-activated cells will come to have a new axis of polarity, oriented at right angles to the original axis. In wounded pea roots, the first indication of this altered polar axis is a reorientation of the cortical MTs which begins within five hours of wounding, long before the cells initiate cell division (Hush *et al.*, 1990). Actin microfilaments also participate in the wound-induced establishment of a new polar axis (Goodbody and Lloyd, 1990)

One manifestation of polarity is the asymmetric distribution of cytoplasmic organelles within the cell. Although this phenomenon has been examined extensively in animals, relatively little work has been done on the mechanism responsible for the distribution and movement of organelles or the morphology of cytomembrane systems in angiosperms. Nevertheless, in many cases the distribution of organelles is not random and there is evidence that the cytoskeleton participates in the spatial orientation of these cytoplasmic structures. For example, statocytes, the gravity perceiving cells of the root cap, are highly polarized cells in which the endoplasmic reticulum (ER) is asymmetrically distributed. A complex of ER cisternae is flattened against the plasma membrane furthest from the apical meristem and a network of cortical MTs lies between these ER cisternae and the plasma membrane. (Sievers and Hensel, 1982). Amyloplasts sediment in a gravitational field and are thought to generate a signal by pressing against this distal ER complex. Drug treatments which destroy the network of cortical MTs underlying the distal complex also result in a loss of the asymmetric distribution of the ER complex. Thus, cortical MTs appear to be responsible for the positioning of the ER cisternae in these cells. Cross-bridges are observed between the MTs and both the plasma membrane and the distal ER complex (Hensel, 1984).

MTs Control the Distribution of Plastids and Other Cellular Structures

Chloroplast distribution has been shown to be non-random in many cells, particularly in algae and vascular cryptograms. Many lower plants have a single large chloroplast per cell. When these cells divide, MTs transport the chloroplast to the pole and participate in the division of the plastid and partitioning of the daughter plastids into the daughter cells (Brown and Lemmon, 1991).

The nucleus itself also is not randomly distributed within the cell, but is positioned by the cytoskeleton, particularly in preparation for cell division. The nucleus is transported from the cell periphery into the center of the cell prior to the division of vacuolate cells. Nuclear migration and the maintenance of this position within the cell prior to division is dependent upon a system of MTs that radiate from the nuclear envelope (Katsuta *et al.*, 1990; Sheldon and Dickinson, 1986; Venverloo and Libbenga, 1987). Nuclear and organelle positioning is apt to be of particular importance during unequal cell divisions. Generative cells of most angiosperms lack plastids so plastid inheritance is strictly maternal. The first division of the male microspore is unequal. The nucleus is transported to one end of the cell prior to division and the plastids are excluded from the daughter cell that becomes the generative cell. Both nuclear migration and plastid distribution within the dividing microspore are dependent on the radial array MTs (Sheldon and Dickinson, 1986; Tanaka, 1991).

Determining the Plane of Cytokinesis

Plant cells do not move about during development As a result, the plane in which they divide will determine the position of the daughter cells within the plant body. Since the position of a cell within the plant ultimately will determine its fate, the plane in which cells divide clearly plays an important role in development. However, it is not the plane in which the mitotic spindle is formed, but the plane in which cytokinesis occurs that determines daughter cell position. Considerable circumstantial evidence indicates that the PPB determines the point at which the growing cell plate meets the parental cell plasma membrane and cell wall (Gunning *et al.*, 1978; Gunning and Wick, 1985; Mineyuki and Gunning, 1990). The problem with this correlation is that there is no obvious mechanism by which the PPB could accomplish this. The PPB MTs are completely gone by metaphase, long before the formation of the phragmoplast which will construct the cell plate. Furthermore, dividing cells do not always form a PPB. Dividing cells lacking a PPB still insert a cell plate into the parental walls during cytokinesis (Mineyuki *et al.*, 1991). Nevertheless, when the PPB is present it accurately predicts the site of cell plate insertion, even in cases where the mitotic spindle forms in a different plane (Palevitz and Hepler, 1974). It has been proposed that the PPB modifies the parental cell wall or plasma membrane so that insertion occurs preferentially at the site so modified, but the nature of these modifications has not been determined. There is evidence that the additional cell wall deposition which occurs beneath the PPB site could serve to mark the point of insertion. Using time-lapse video microscopy, Mineyuki and Gunning (1990) noted that the cell plate, which was fluid and wrinkled before insertion, became rigid and flat after insertion at the PPB-marked site. However, when the site of insertion was experimentally altered so that it did not occur at the PPB-marked site, plate flattening did not occur and the cross wall formed was abnormal. Although the PPB MTs are gone long before cytokinesis, actin filaments, which were formed while the PPB was present, connect the spindle and the PPB site. These actin filaments persist through mitosis and they could guide the phragmoplast to the site of insertion (Lloyd and Traas, 1988).

Determination of Cell Shape

Microtubules have a powerful and direct effect on cell shape in organisms lacking cell walls. Heliozoans, for example, have numerous needle-like axopodia which project from their surface. These contain several hundred bundled axonemal MTs that are responsible for extending and retracting the axopodia as the organisms moves about and contacts the substratum (Ando and Shigenaka, 1989). In contrast, plant cell shape is determined primarily by the cell wall and by physical forces generated through cell-cell contacts (Hardham *et al.*, 1980). This can be demonstrated by enzymatically removing the cell wall. In most cases the resultant protoplasts assume a spherical shape, regardless of the shape of the cells from which the protoplasts were derived (Lloyd *et al.*, 1980). However, there is evidence that MTs can play a limited direct role in determining cell shape, and there is at least one cell type, the generative cell, in which MTs are the most significant direct determinant of cell shape. Although protoplasts usually are spherical, the conditions under which they are

contract as a result of a lateral expansion of the cell walls, a shift in the orientation of the cortical MTs precedes this lateral expansion (Cyr *et al.*, 1988).

Cortical MTs also have been implicated in determining the orientation of cellulose MF during the deposition of patterned secondary walls in differentiation tracheary elements. Certainly drug treatments that destroy cortical MTs repeatedly have been shown to disrupt the pattern of secondary wall deposition (Hepler and Fosket, 1971; Hogetsu, 1991; Pickett-Heaps, 1967). However, in some cases, the alignment of cellulose MF may be controlled by additional factors, other than cortical MTs, which may play a dominant role in orienting the deposition of MF in some cell types (Preston, 1988; Wilms *et al.*, 1990). Also, cortical MTs do not appear to play any role in the growth or shaping of the pollen tube (Steer and Steer, 1989).

The mechanism by which MTs control the orientation of cellulose MF deposition is not known. However, there is evidence that the cellulose MF deposition occurs via plasma membrane-bound cellulose synthase complexes, which have been visualized by electron microscopy, although they have not been characterized biochemically (Giddings and Staehelin, 1988; Giddings *et al.*, 1980). In the various models that have been put forward to account for the guiding role of MTs in MF deposition, crossbridges are assumed to occur between either the MTs and the cellulose synthase complexes directly, or between the integral membrane proteins other than the cellulose synthase complexes and the cortical MTs (Fig. 5) (Heath and Seagull, 1982; Giddings and Staehelin, 1988). In the former case, the crosslink between cortical MTs and the cellulose synthase complex could be a mechanoenzyme, similar to kinesin, which would transport the complex through the membrane along the microtubule. In the latter case, the cross bridges between integral membrane proteins and MTs would create channels through which the synthase complex could move preferentially, but there would be no direct connection between the complexes and the MTs.

Also unexplained is the nature of the factors controlling the orientation of the cortical MTs. Hormones can influence MT orientation. Ethylene treatment of root or stem tissue causes the growing tissues to expand laterally and this is preceded by the switch in MT orientation from transverse to random (Apelbaum and Burg, 1971; Lang *et al.*, 1982; Roberts *et al.*, 1985). Gibberellic acid treatment conversely promotes the transverse orientation of cortical MTs, as well as cell elongation, in various tissues (Akashi and Shibaoka, 1987; Mita and Shibaoka, 1984a,b).

Furthermore, in some cases these effects can be reversed by abscisic acid (Sakiyama and Shibaoka, 1990). Light also has been shown to alter the orientation of cortical MTs in some organs in which light affects the elongation rate (Iwata and Hogetsu, 1989a; Murata and Wada, 1989). It seems unlikely that hormones or light directly control MT orientation. Rather, these agents may alter some aspect of cell physiology which in turn regulates MT orientation. Because of the intimate physical connection between integral plasma membrane proteins and cortical MTs (Akashi and Shibaoka, 1991), these effects could be mediated through the plasma membrane. Extracellular ionic currents have been detected around pea roots and these currents are affected by wounds which also result in the reorientation of cortical MTs (Hush and Overall, 1989). Lloyd has suggested that the cortical MTs become transversely oriented through extensive cross-bridging (Lloyd, 1984). An increase in cross-bridging would tend to align cortical MTs, and even to bring them into a transverse orienta-

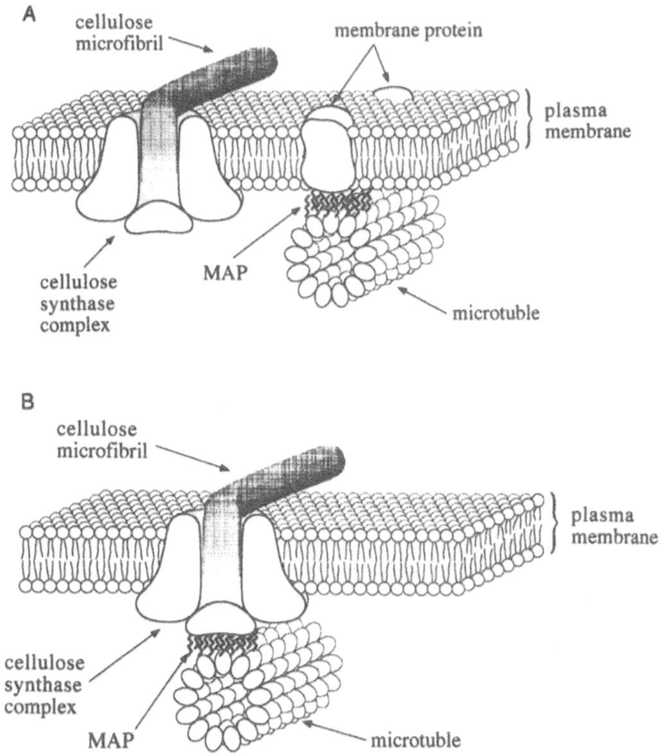

Figure 5. Models for mechanisms by which cortical MTs could control the orientation of cellulose MFs. Cellulose synthase, a complex of integral membrane proteins, synthesizes cellulose MFs on the outer surface of the cell from precursors provided from within the cell. (A) Cross-bridging proteins between cellulose synthase complexes and cortical MTs may either restrict or guide the movement of the complex. If the cross-bridging protein were a mechanoenzyme, such as kinesin, the complex could be actively transported along the MT. (B) The MT may be crossbridged to other integral membrane proteins which create channels through which the synthase complex may move (Re-drawn from Giddings and Staehelin, 1988).

tion. Certainly there are cross-bridges between cortical MTs, but whether these increase as the cortical array assumes a transverse orientation is not known. Since the crossbridges most likely are MAPs, the gene encoding these MAPs could be regulated by hormones. Modulation of the cellular levels of MAP proteins then would change the extent of cortical MT cross-bridging, leading to a greater or lesser degree of transverse orientation of the array. Alternatively, the interaction of cortical MTs with crossbridging MAPs could be regulated by the tubulin isotypes from which they are assembled. The incorporation of some tubulin isotypes into the MTs of the cortical array may permit a more stable interaction of the MAP with the MTs.

MICROTUBULE FUNCTION IS MEDIATED BY MICROTUBULE-ASSOCIATED PROTEINS

Purified tubulin dimers are capable of self-assembling *in vitro* to form MTs in the absence of other proteins (Lee, 1982). Nevertheless, microtubules assembled from crude cell homogenates contain heterogeneous assortment of co-polymerizing proteins, in addition to tubulin, some of which have been shown to be MAPs (Matus, 1988). Electron microscopy shows MAPs as linear projections from MTs, forming cross-bridges between MTs, membranes and other cellular structures, including the plasma membrane and cytoplasmic vesicles (Cooper *et al.*, 1990; Hirokawa *et al.*, 1988a,b; Meluh and Rose, 1990). Biochemically, MAPs are proteins that bind to assembled MTs and promote MT assembly. MAPs are not structural parts of MTs, but they may determine most, if not all, of the cellular functions of MTs. Motility and transport functions of MTs are brought about by MT-dependent ATPases which convert chemical energy to mechanical work. These mechanoenzymes include the kinesins and dynein (Scholey, 1990; Schroer *et al.*, 1988). Kinesins appear to be a relatively large family of related proteins. The *Drosophila* genome may have as many as 30 genes encoding kinesin-like proteins (Endow and Hatsumi, 1991). While MAPs, such as kinesin and dynein, function as molecular motors, others such as *tau* can promote MT assembly or stabilize MTs to depolymerization (Drubin and Kirschner, 1986; Lee, 1990). Angiosperm MAPs are not well characterized at the present time, although a MT bundling protein was isolated from carrot (Cyr and Palevitz, 1989) and a maize MAP was shown to be immunologically related to the brain MAP *tau* (Vantard *et al.*, 1991). The well characterized MAPs, such as *tau* and MAP2, contain a conserved MT binding domain consisting of three or four imperfect repeats of an 18 amino acid sequence (Aizawa *et al.*, 1988; Goedert *et al.*, 1991). Plant MAPs could be expected to contain similar MT binding domains. However, the remainder of the molecule may be considerably diverged from vertebrate MAPs.

THE EXPRESSION OF SOME TUBULIN GENES IS DEVELOPMENTALLY REGULATED

The individually expressed vertebrate tubulin genes exhibit varying degrees of developmental regulation (Sullivan, 1988). For example, one diverged β–tubulin gene in chickens is expressed only in erythrocytes and thrombocytes and its product is the only β–tubulin present in the MTs of the erythrocyte marginal band (Murphy *et al.*, 1987). Mammals have a comparably expressed β–tubulin gene that encodes a β–tubulin isotype utilized for the formation of erythroblast marginal band MTs (Wang *et al.*, 1986). Attempts to demonstrate that the tubulins encoded by these genes are highly specialized for marginal band function have so far failed. The erythropoetic-specific β–tubulin was incorporated into other MT arrays when this gene was transfected into fibroblast, demonstrating that it is not so specialized that it cannot be incorporated into other MT arrays (Lewis *et al.*, 1987). Other developmentally expressed vertebrate tubulins also can assemble into various MT arrays and function in those arrays.

The expression of several angiosperm tubulin genes is developmentally regulated. Two very similar maize α-tubulin genes are expressed primarily in the radicle (Montoliu *et al.*, 1990b), while a third maize α-tubulin gene is not expressed preferentially in any organ, although transcripts from the latter gene are most abundant in dividing cells (Montoliu *et al.*, 1990a). The *Arabidopsis* β-1-tubulin gene is expressed preferentially in roots (Oppenheimer *et al.*, 1988), while the *Arabidopsis* α-1 tubulin gene is expressed almost exclusively in flowers (Ludwig *et al.*, 1988). A chimeric β-glucuronidase gene, when fused to the *Arabidopsis* α-1 tubulin gene promoter, was expressed almost exclusively in the pollen of transgenic plants (Carpenter *et al.*, 1992). The β-1-tubulin of soybean is expressed strongly only in the hypocotyl of etiolated seedlings and this expression is down-regulated by light (Han *et al.*, 1991). At present there are no reports of MT array-specific tubulin isotypes in angiosperms.

FUNCTIONAL SPECIALIZATION OF TUBULIN ISOTYPES

The significance of tubulin isotypes and isoforms is not known. It has been proposed that tubulin isotypes are differentiated functionally so that each is specialized for a different MT array. This is known as the "multitubulin hypothesis" (Fulton and Simpson, 1976). A strict interpretation of the multitubulin hypothesis postulates that each tubulin isotype performs an essential function in a particular MT array for which other tubulin isotypes cannot substitute. It also implies the existence of some sorting mechanism by which different isotypes are used to construct different MT arrays. This hypothesis clearly is wrong as a general statement. Several different α- and β-tubulin isotypes are expressed in HeLa cells, 3T3 cells and cultured carrot cells. These mingle freely and are used at random to construct all the MT arrays found in these cells (Dawson and Lloyd, 1985; Lewis *et al.*.1987). Some cell types express only or predominately a single α- or β-tubulin isotype, but in cells that express multiple tubulin isotypes there is no evidence for the sorting of isotypes to different MT arrays. However, on the question of the possible functional specialization of tubulin isotypes, there is evidence that both supports and rejects the multitubulin hypothesis, depending upon the organism examined. Tubulins are encoded by small multigene families in some fungi. In both *Aspergillus* and yeast, the tubulins encoded by the different members of these β-tubulin multigene families are interchangeable, even though they exhibit considerable sequence diversity and their expression is regulated developmentally (May, 1989; May *et al.*, 1985; Schatz *et al.*, 1986a; Schatz *et al.*, 1986b).

On the other hand, the *Drosophila* β-3-tubulin isotype has been demonstrated to be functionally specialized, while there is strong evidence that a neuron-specific β-tubulin isotype in the nematode *Caenorhabditis elegans* is specialized structurally and functionally. *Drosophila* has four α- and four β-tubulin genes, some of which are expressed constitutively, while the expression of others is regulated developmentally (Raff, 1984). The β-2-tubulin gene is expressed exclusively in the testis where its product is the most abundant β-tubulin. Beta-2-tubulin is utilized indiscriminately for all the MT arrays of the testis, including the mitotic and meiotic spindles, a MT array known as the manchette that is involved in nuclear shaping during spermatogenesis, and the MTs of the sperm flagellar axoneme

(Kopczak *et al.*, 1992). Also, three of the *Arabidopsis* α–tubulins have a lysine residue at position 40 whose epsilon-amino group could be acetylated, while the other three have a different amino acid at this position so that they could not be posttranslationally modified by this mechanism (Kopczak *et al.*, 1992). *Chlamydomonas* flagellar axonemes contain an α–tubulin isoform that arises through posttranslational acetylation of the lysine at position 40 in one of the two α–tubulin gene products (L' Hernault and Rosenbaum, 1985b). Alpha-tubulin is acetylated when the tubulin dimers assemble into MTs of the flagellar axoneme and it is deacetylated during disassembly (L' Hernault and Rosenbaum, 1985a). Flagellar axoneme MTs are more stable than the cytoplasmic MTs of this organism. Similarly, where acetylated tubulin also occurs in cytoplasmic MTs, those MTs containing the acetylated tubulin are more stable (Webster and Borisy, 1989). When tubulin is acetylated *in vitro*, this posttranslational modification has no effect on the stability of MTs assembled from the modified tubulin (Maruta *et al.*, 1986). As a result, it is likely that acetylation, and perhaps all other posttranslational modifications of tubulins, affect MT stability indirectly by marking sets of MTs for interaction with MAPs (Billger *et al.*, 1991).

ACKNOWLEDGMENTS

The preparation of this manuscript and the research described herein that was conducted in the laboratory of D.E.F. was supported by a grant from the National Science Foundation (DCB-9004693) and by a grant from the Shimadzu Corporation, Kyoto, Japan to D.E.F.

REFERENCES

Aizawa H, Kawasaki H, Murofushi H, Kotani S, Suzuki K, Sakai H (1988) Microtubule-binding domain of *tau* proteins. *J Biol Chem* 263: 7703-7707.

Akashi T, Shibaoka H (1987) Effects of gibberellin on the arrangement and the cold stability of cortical microtubules in epidermal cells of pea internodes. *Plant Cell Physiol* 28: 339-348.

Akashi T, Shibaoka H (1991) Involvement of transmembrane proteins in the association of cortical microtubules with the plasma membrane in tobacco BY-2 cells. *J Cell Sci* 98: 169-174.

Akashi T, Izumi K, Nagano E, Enomoto M, Mizuno K, Shibaoka H (1988) Effects of propyzamide on tobacco cell microtubules *in vivo* and *in vitro*. *Plant Cell Physiol* 29: 1053-1062.

Ando M, Shigenaka Y (1989) Structure and function of the cytoskeleton in heliozoa: I. Mechanism of rapid axopodial contraction in *Echinosphaerium*. *Cell Motil Cytoskeleton* 14: 288-301.

Apelbaum A, Burg SP (1971) Altered cell microfibrillar orientation in ethylene-treated *Pisum sativum* stems. *Plant Physiol* 48: 648-652.

Aya M, Takase I, Kishinno S, Kurihara K (1975) Amiprophosmethyl, a new herbicide in upland crops. *In*: Proceedings of Fifth Asian-Pacific Weed Society Conference, Tokyo, Japan, pp 138-141.

Bajer AS, Cypher C, Mole-Bajer J, Howard HM (1982) Taxol-induced anaphase reversal: evidence that elongating microtubules can exert a pushing force in living cells. *Proc Natl Acad Sci USA* 79: 6569-6573.

Bajer AS, Mole-Bajer J (1986) Drugs with colchicine-like effects that specifically disassemble plant but not animal microtubules. *In*: Dynamic Aspects of Microtubule Biology, Soifer D (ed), New York Academy of Sciences, New York, pp 767-784.

Baskin TI, Cande WZ (1990) The structure and function of the mitotic spindle in flowering plants. *Annu Rev Plant Physiol Plant Mol Biol* 41: 277-315.

Baum SG, Wittner M, Nadler JP, Horwitz SB, Dennis J, Schiff PB, Tanowitz HB (1981) Taxol, a microtubule stabilizing agent, blocks the replication of *Trypanosoma cruzi*. *Proc Natl Acad Sci USA* 78: 4571-4575.

Bayley PM, Manser EJ (1985) Assembly of microtubules from nucleotide-depleted tubulin. *Nature* 318: 683-685.

Beese L, Stubbs G, Cohen C (1987) Microtubule structure at 18 Å resolution. *J Mol Biol* 194: 257-264.

Billger M, Str mberg E, Wallin M (1991) Microtubule-associated proteins-dependent colchicine stability of acetylated cold-labile brain microtubules from the Atlantic cod, *Gadus morhua*. *J Cell Biol* 113: 331-338.

Blakeslee AF, Avery A (1937) Methods of inducing doubling of chromosomes in plants. *J Hered* 28: 393-411.

Borisy GG, Marcum JM, Olmsted JB, Murphy DB, Johnson KA (1975) Purification of tubulin and associated high molecular weight proteins from porcine brain and characterization of microtubule assembly *in vitro*. *Ann NY Acad Sci* 253: 107-132.

Brown RC, Lemmon BE (1990) Monoplastidic cell division in lower land plants. *Amer J Bot* 77: 559-571.

Brown RC, Lemmon BE (1991) Plastid polarity and meiotic spindle development in microsporogenesis of *Selaginella*. *Protoplasma* 161: 168-180.

Burns R (1987) Tubulin's terminal tyrosine. *Nature* 327: 103-104.

Burns RG (1991) α–, β–, and γ–tubulins: Sequence comparisons and structural constraints. *Cell Motil Cytoskeleton* 20: 181-189.

Busby CH, Gunning BES (1989) Development of the quadripolar meiotic apparatus in *Funaria* spore mother cells: Analysis by means of anti-microtubule drug treatments. *J Cell Sci* 93: 267-277.

Calarco-Gillam PD, Siebert MC, Hubble R, Mitchison T, Kirschner M (1983) Centrosome development in early mouse embryos as defined by an auto-antibody against pericentriolar material. *Cell* 35: 621-629.

Carpenter JL, Ploense SE, Snustad DP, Silflow CD (1992) Preferential expression of an α–tubulin gene of *Arabidopsis* in pollen. *Plant Cell* 4: 557-571.

Cassimeris L, Pryer NK, Salmon ED (1988) Real-time observations of microtubule dynamic instability in living cells. *J Cell Biol* 107: 2223-2231.

Centonze VE, Borisy GG (1990) Nucleation of microtubules from mitotic centrosomes is modulated by a phosphorylated epitope. *J Cell Sci* 95: 405-411.

Clayton L, Black CM, Lloyd CW (1985) Microtubule nucleating sites in higher plant cells identified by an auto-antibody against pericentriolar material. *J Cell Biol* 101: 319-324.

Cleary AL, Hardham AR (1990) Reinstatement of microtubule arrays from cortical nucleating sites in stomatal complexes of *Lolium rigidum* following depolymerization of microtubules by oryzalin and high pressure. *Plant Cell Physiol* 31: 903-915.

Cooper MS, Cornell-Bell AH, Chernjavsky A, Dani JW, Smith SJ (1990) Tubulovesicular processes emerge from *trans*-Golgi cisternae, extend along microtubules, and interlink adjacent *trans*-Golgi elements into a reticulum. *Cell* 61: 135-145.

Cowan NJ, Lewis SA, Gu W, Burgoyne RD (1988) Tubulin isotypes and their interaction with microtubule associated proteins. *Protoplasma* 145: 106-111.

Cross D, Dominguez J, Maccioni RB, Avila J (1991) MAP-1 and MAP-2 binding sites at the *C*–terminus of β–tubulin. Studies with synthetic tubulin peptides. *Biochemistry* 30: 4362-4366.

Cyr RJ, Lin B-L, Jernstedt JA (1988) Root contraction in hyacinth. II. Changes in tubulin levels, microtubule number and orientation associated with differential cell cxpansion. *Planta* 174: 446-452.

Cyr RJ, Palevitz BA (1989) Microtubule-binding proteins from carrot. I. Initial characterization and microtubule bundling. *Planta* 177: 245-260.

Dawson PJ, Lloyd CW (1985) Identification of multiple tubulins in taxol microtubules purified from carrot suspension cells. *EMBO* J 4: 2451-2455.

De Brabander M, Geuens G, Nuydens R, Willebrords R, De Mey J (1981) Microtubule assembly in living cells after release from nocodazole block: the effects of metabolic inhibitors, taxol and pH. *Cell Biol Int Rep* 5: 913-920

Deysson G (1975) Microtubules and antimitotic substances. *In*: Microtubules and Microtubule Inhibitors, Borgers M, deBrabander M (eds), North-Holland, Amsterdam, pp 427-451.

Drubin DG, Kirschner MW (1986) *Tau* protein function in living cells. *J Cell Biol* 103: 2739-2746.

Edde B, Rossier J, Le Caer J-P, Desbruyeres E, Gros F, Denoulet P (1990) Posttranslational glutamylation of α–tubulin. *Science* 247: 83-85.

Eigsti OJ, Dustin Jr. (1955) Colchicine in Agriculture, Medicine, Biology and Chemistry. Iowa State Univ Press, Ames, Iowa.

Endow SA, Hatsumi M (1991) A multimember kinesin gene family in *Drosophila*. *Proc Natl Acad Sci USA* 88: 4424-4427.

Flanders DJ, Rawlins DJ, Shaw PJ, Lloyd CW (1990) Nucleus-associated microtubules help determine the division plane of plant epidermal cells: Avoidance of four-way junctions and the role of cell geometry. *J Cell Biol* 110: 1111-1122.

Fosket DE (1989) Cytoskeletal proteins and their genes in higher plants. *In*: The Biochemistry of Plants, Stumpf P, Conn EE (eds), vol. 15: Molecular Biology, Academic Press, New York, pp 392-454.

Fosket DE, Morejohn LC (1992) Structural and functional organization of tubulin. *Annu Rev Plant Physiol Plant Mol Biol* 43: 201-240.

Frohman MA (1990) RACE: Rapid amplification of cDNA ends. *In*: PCR Protocols: A Guide to Methods and Applications, Innis MA, Gelfand DH, Sninsky JJ, White TJ (eds), Academic Press, San Diego, pp 28-38.

Fulton C, Simpson PA (1976) Selective synthesis and utilization of flagellar tubulin: The multitubulin hypothesis. *In*: Cell Motility, Goldman R, Polland T, Rosenbaum J (eds), Cold Springs Harbor Lab., New York, pp 987-1005.

Galatis B, Apostolakos P, Katsaros CHR, Loukari H (1982) Pre-prophase microtubule band and local wall thickening in guard cell mother cells of some *Leguminosae*. *Ann Bot* 50: 779-791.

Gasch A, Hinz U, Leiss D, Renkawitz-Pohl R (1988) The expression of b1 and b3-tubulin genes of *Drosophila melanogaster* is spatially regulated during embryogenesis. *Mol Gen Genet* 211: 8-16.

Giddings TH, Jr Staehelin LA (1988) Spatial relationship between microtubules and plasma-membrane rosettes during the deposition of primary wall microfibrils in *Closterium* sp. *Planta* 173: 22-30.

Giddings TH, Jr, Brower DL, Staehelin LA (1980) Visualization of particle complexes in the plasma membrane of *Micrasterias denticulata* associated with the formation of cellulose fibrils in primary and secondary cell walls. *J Cell Biol* 84: 327-339.

Goedert M, Crowther RA, Garner CC (1991) Molecular characterization of microtubule-associated proteins *tau* and MAP2. *Trends Neurosci.* 14: 193-199.

Goodbody KC, Lloyd CW (1990) Actin filaments line up across *Tradescantia* epidermal cells, anticipating wound-induced division planes. *Protoplasma* 157: 92-101.

Gould RR, Borisy GG (1977) The pericentriolar material in Chinese hamster ovary cells nucleates microtubule formation. *J Cell Biol* 73: 601-615.

Green PB (1983) Analysis of axis extension. *In*: Positional Controls in Plant Development, Barlow PW, Carr DJ (eds), Cambridge University Press, pp 53-82.

Guiltinan MJ, Ma D-P, Barker RF, Bustos MM, Cyr RJ, Yadegari R, Fosket.DE (1987) The isolation, characterization and sequence of two divergent β–tubulin genes from soybean (*Glycine max* L.). *Plant Mol Biol* 10: 171-184.

Gunning BES, Wick SM (1985) Preprophase bands, phragmoplasts, and spatial control of cytokinesis. *J Cell Sci Suppl* 2: 157-179.

Gunning BES, Sammut M (1990) Rearrangements of microtubules involved in establishing cell division planes start immediately after DNA synthesis and are completed just before mitosis. *Plant Cell* 2: 1273-1282.

Gunning BES, Hardham AR, Hughes JE (1978) Pre-prophase bands of microtubules in all categories of formative and proliferative cell divisions in *Azolla* roots. *Planta* 143: 145-160.

Hains FO, Dickerson RM, Wilson L, Owellen RJ (1978) Differences in the binding properties of vinca alkaloids and colchicine to tubulin by varying proteins sources and methodology. *Biochem Pharmacol* 27: 71-76.

Han I-S, Jongewaard I, Fosket DE (1991) Limited expression of a diverged β–tubulin gene during soybean (*Glycine max* [L.] Merr.) development. *Plant Mol Biol* 16: 225-234.

Hardham AR, Gunning BES (1978) Structure of cortical microtubule arrays in plant cells. *J Cell Biol* 77: 14-33.

Hardham AR, Green PB, Lang JM (1980) Reorganization of cortical microtubules and cellulose deposition during leaf formation in *Graptopetalum paraguayense*. *Planta* 149: 181-195.

Harper JDI, Mitchison JM, Williamson RE, John PCL (1989) Does the autoimmune serum 5051 specifically recognise microtubule organising centres in plant cells. *Cell Biol Int Rep* 13: 471-484.

Hasezawa S, Marc J, Palevitz BA (1991) Microtubule reorganization during the cell cycle in synchronized BY-2 tobacco suspensions. *Cell Motil Cytoskeleton* 18: 94-106.

Heath IB, Seagull RW (1982) Oriented cellulose fibrils and the cytoskeleton: A critical comparison of models. *In*: The Cytoskeleton in Plant Growth and Development, Lloyd CW (ed), Academic Press, New York, pp 163-181.

Hensel W (1984) A role of microtubules in the polarity of statocytes from roots of *Lepidium sativum* L. *Planta* 162: 404-414.

Hepler PK, Fosket DE (1971) The role of microtubules in vessel member differentiation in *Coleus*. *Protoplasma* 72: 213-236.

Hepler PK, Bonsignore CL (1990) Caffeine inhibition of cytokinesis: Ultrastructure of cell plate formation/degradation. *Protoplasma* 157: 182-192.

Hepler PK, Jackson WT (1968) Microtubules and early stages of cell-plate formation in the endosperms of *Haemanthus katherinae* Baker. *J Cell Biol* 38: 437-446.

Hirokawa N, Hisanaga S, Shiomura Y (1988a) MAP2 is a component of crossbridges between microtubules and neurofilaments in the neuronal cytoskeleton: Quick-freeze, dep-etch immunoelectron microscopy and reconstitution studies. *J Neurosci* 8: 2769-2779.

Hirokawa N, Shiomura Y, Okabe S (1988b) *Tau* proteins: The molecular structure and mode of binding on microtubules. *J Cell Biol* 107: 1449-1459.

Hogan CJ, Cande WZ (1990) Antiparallel microtubule interactions: Spindle formation and anaphase B. *Cell Motil Cytoskeleton* 16: 99-103.

Hogetsu T (1991) Mechanism for formation of the secondary wall thickening in tracheary elements:

Maruta H, Greer K, Rosenbaum JL (1986) The acetylation of α–tubulin and its relationship to the assembly and disassembly of microtubules. *J Cell Biol* 103: 571-579.

Matus A (1988) Microtubule-associated proteins: Their potential role in determining neuronal morphology. *Annu Rev Neurosci* 11: 29-44.

May GS (1989) The highly divergent β–tubulins of *Aspergillus nidulans* are functionally interchangeable. *J Cell Biol* 109: 2267-2274.

May GS, Gambino J, Weatherbee JA, Morris NR (1985) Identification and functional analysis of β–tubulin genes by site specific integrative transformation in *Aspergillus nidulans*. *J Cell Biol* 101: 712-719.

Melan MA (1990) Taxol maintains organized microtubule patterns in protoplasts which lead to the resynthesis of organized cell wall microfibrils. *Protoplasma* 153: 169-177.

Meluh PB, Rose MD (1990) KAR3, a kinesin-related gene required for yeast nuclear fusion. *Cell* 60: 1029-1041.

Mineyuki Y, Gunning BES (1990) A role for preprophase bands of microtubules in maturation of new cell walls, and a general proposal on the function of preprophase band sites in cell division in higher plants. *J Cell Sci* 97: 527-537.

Mineyuki Y, Murata T, Wada M (1991) Experimental obliteration of the preprophase band alters the site of cell division, cell plate orientation and phragmoplast expansion in *Adiantum* protonemata. *J Cell Sci* 100: 551-557.

Mita T, Shibaoka H (1984a) Effects of S-3307, an inhibitor of gibberellin biosynthesis, on swelling of leaf sheath cells and on the arrangement of cortical microtubules in onion seedlings. *Plant Cell Physiol* 25: 1531-1539.

Mita T, Shibaoka H (1984b) Gibberellin stabilizes microtubules in onion leaf sheath cells. *Protoplasma* 119: 100-109.

Mitchison TJ, Kirschner MW (1987) Some thoughts on the partitioning of tubulin between monomer and polymer under conditions of dynamic instability. *Cell Biophys* 11: 35-55.

Monteiro MJ, Cleveland DW (1988) Sequence of chicken cβ7 tubulin. Analysis of a complete set of vertebrate β–tubulin isotypes. *J Mol Biol* 199: 439-446.

Montoliu L, Puigdomenech P, Rigau J (1990a) The Tuba3 gene from *Zea mays*: Structure and expression in dividing plant tissues. *Gene* 94: 201-207.

Montoliu L, Rigau J, Puigdomenech P (1990b) A tandem of α–tubulin genes preferentially expressed in radicular tissues from *Zea mays*. *Plant Mol Biol* 14: 1-15.

Morejohn LC, Fosket DE (1984a) Taxol-induced rose microtubule polymerization *in vitro* and its inhibition by colchicine. *J Cell Biol* 99: 141-147.

Morejohn LC, Fosket DE (1984b) Inhibition of plant microtubule polymerization *in vitro* by the phosphoric amide herbicide amiprophos-methyl. *Science* 224: 874-876.

Morejohn LC, Fosket DE (1992) The biochemistry of compounds with anti-microtubule activity in plant cells. *Pharmac Ther* 51: 217-230.

Morejohn LC, Bureau TE, Tocchi LP, Fosket DE (1984) Tubulins from different higher plant species are immunologically nonidentical and bind colchicine differentially. *Proc Natl Acad Sci USA* 81: 1440-1444.

Morejohn LC, Bureau TE, Mole-Bajer J, Bajer AS, Fosket DE (1987a) Oryzalin, a dinitroaniline herbicide, binds to plant tubulin and inhibits microtubule polymerization *in vitro*. *Planta* 1-13.

Morejohn LC, Bureau TE, Tocchi LP, Fosket DE (1987b) Resistance of *Rosa* microtubule polymerization to colchicine results from a low-affinity interaction of colchicine and tubulin. *Planta* 170: 230-241.

Murata T, Wada M (1989) Organization of cortical microtubules and microfibril deposition in response to blue-light-induced apical swelling in a tip-growing *Adiantum* protonema cell. *Planta* 178: 334-341.

Murphy DB, Wallis KT, Machlin PS, Ratrie H III, Cleveland DW (1987) The sequence and expression of the divergent β–tubulin in chicken erythrocytes. *J Biol Chem* 262: 14305-14312.

Oakley BR, Oakley CE, Yoon Y, Jung MK (1990) γ–tubulin is a component of the spindle pole body that is essential for microtubule function in *Aspergillus nidulans*. *Cell* 61: 1289-1301.

Oppenheimer DG, Haas N, Silflow CD, Snustad DP (1988) The β–tubulin gene family of *Arabidopsis thaliana*: preferential accumulation of the b1 transcript in roots. *Gene* 63: 87-102.

Palevitz BA (1987) Accumulation of F-actin during cytokinesis in *Allium*. Correlation with microtubule distributuin and the effects of drugs. *Protoplasma* 141: 24-32.

Palevitz BA, Hepler PK (1974) Unique cell division and cytodifferentiation The control of the plane of division during stomatal differentiation in *Allium* I. Spindle reorientation. *Chromosoma* 46: 297-326.

Parness J, Horwitz SB (1981) Taxol binds to polymerized tubulin *in vitro*. *J Cell Biol* 91: 479-487.

Parthasarathy MV, Perdue TD, Witztum A, Alvernaz J (1985) Actin network as a normal component of the cytoskeleton in many vascular plant cells. *Amer J Bot* 72: 1318-1323.

Pickett-Heaps JD (1967) The effects of colchicine on the ultrastructure of dividing plant cells, xylem wall differentiation and distribution of cytoplasmic microtubules. *Dev Biol* 15: 206-236.

Pickett-Heaps JD (1969) The evolution of the mitotic apparatus: an attempt at comparative ultrastructural cytology in dividing plant cells. *Cytobios* 3: 257-280.

Prescott AR, Foster KE, Warn RM, Gull K (1989) Incorporation of tubulin from an evolutionarily diverse source, *Physarum polycephalum*, into the microtubules of a mammalian cell. *J Cell Sci* 92: 595-605.

Preston RD (1988) Cellulose-microfibril-orienting mechanisms in plant cell walls. *Planta* 174: 67-74.

Quatrano R (1978) Development of cell polarity. *Annu Rev Plant Physiol* 29: 487-510.

Raff EC (1984) Genetics of microtubule systems. *J Cell Biol* 99: 1-10.

Rivas CI, Vera JC, Maccioni RB (1988) Anti-idiotypic antibodies that react with microtubule-associated proteins are present in the sera of rabbits immunized with synthetic peptides from tubulin s regulatory domain. *Proc Natl Acad Sci USA* 85: 6092-6096.

Roberts IN, Lloyd CW, Roberts K (1985) Ethylene-induced microtubule reorientations: mediation by helical arrays. *Planta* 164: 439-447.

Robinson DG, Herzog W (1977) Structure, synthesis and orientation of microfibrils. III. A survey of the action of microtubule inhibitors on microtubules and microfibril orientation in *Oocystis solitaria*. *Cytobiologie* 15: 463-484.

Robinson DG, Grimm I, Sachs H (1976) Colchicine and microfibril orientation. *Protoplasma* 89: 375-380.

Rudolph JE, Kimble M, Hoyle HD, Subler MA, Raff EC (1987) Three *Drosophila* β–tubulin sequences: a developmentally regulated isoform (b3), the testis-specific isoform (b2), and an assembly-defective mutation of the testis-specific isoform (b2t8) reveal both an ancient divergence in metazoan isotypes and structural constraints for β–tubulin function. *Mol Cell Biol* 7: 2231-2242.

Sackett DL, Lippoldt RE (1991) Thermodynamics of reversible monomer-dimer association of tubulin. *Biochemistry* 30: 3511-3517.

Sakiyama M, Shibaoka H (1990) Effects of abscisic acid on the orientation and cold stability of cortical microtubules in epicotyl cells of the dwarf pea. *Protoplasma* 157: 165-171.

Savage C, Hamelin M, Culotti JG, Coulson A, Albertson DG, Chalfie M (1989) *mec*-7 is a β–tubulin gene required for the production of 15-protofilament microtubules in *Caenorhabditis elegans*. *Genes Dev* 3: 870-881.

Savage C, Chalfie M (1991) Genetic aspects of microtubule biology in the nematode *Caenorhabditis elegans. Cell Motil Cytoskeleton* 18: 159-163.

Schatz PJ, Pillus L, Grisafi P, Solomon F, Botstein D (1986a) Two functional α–tubulin genes of the yeast *Saccharomyces cerevisiae* encode divergent proteins. *Mol Cell Biol* 6: 3711-3721.

Schatz PJ, Solomon F, Botstein D (1986b) Genetically essential and nonessential α–tubulin genes specify functionally interchangeable proteins. *Mol Cell Biol* 6: 3722-3733.

Scheele RB, Bergen LG, Borisy GG (1982) Control of the structural fidelity of microtubules by initiation sites. *J Mol Biol* 154: 485-500.

Schiff PB, Fant J, Horwitz SB (1979) Promotion of microtubule assembly *in vitro* by taxol. *Nature* 277: 665-667.

Schiff PB, Horowitz SB (1981) Taxol assembles tubulin in the absence of exogenous guanosine 5'-triphosphate or microtubule associated proteins. *Biochemistry* 20: 3247-3251.

Schnepf E (1986) Cellular Polarity. *Annu Rev Plant Physiol* 37: 23-47.

Scholey JM (1990) Cell motility: Multiple microtubule motors. *Nature* 343: 118-120.

Schroer TA, Schnapp BJ, Reese TS, Sheetz MP (1988) The role of kinesin and other soluble factors in organelle movement along microtubules. *J Cell Biol* 107: 1785-1792.

Seagull RW, Falconer MM, Weerdenburg CA (1987) Microfilaments: dynamic arrays in higher plant cells. *J Cell Biol* 104: 995-1004.

Sheldon JM, Dickinson HG (1986) Pollen wall formation in *Lilium*:the effect of chaotropic agents, and the organisation of the microtubular cytoskeleton during pattern development. *Planta* 168: 11-23.

Sievers A, Hensel W (1982) The nature of graviperception. *In*: Plant Growth Substances, Wareing PF (ed), Academic Press, London, pp 497-506.

Silflow CD, Oppenheimer DG, Kopczak SD, Ploense SE, Ludwig SR, Haas N, Snustad DP (1987) Plant tubulin genes: Structure and differential expression during development. *Dev Genet* 8: 435-460.

Sinnott EW, Bloch R (1945) The cytoplasmic basis of intercellular patterns in vascular differentiation. *Amer J Bot* 32: 151-156.

Snustad DP, Haas NA, Kopczak SD, Silflow CD (1992) The small genome of *Arabidopsis* contains at least nine expressed β–tubulin genes. *Plant Cell* 4: 549-556.

Stearns T, Evans L, Kirschner M (1991) γ–tubulin is a highly conserved component of the centrosome. *Cell* 65: 825-836.

Steer MW, Steer JM (1989) Pollen tube tip growth. *New Phytol* 111: 323-358.

Strachan SD, Hess FD (1983) The biochemical mechanism of action of the dinitroaniline herbicide oryzalin. *Pest Biochem Physiol* 20: 1-10.

Sullivan KF (1988) Structure and utilization of tubulin isotypes. *Annu Rev Cell Biol* 4: 687-716.

Tanaka I (1991) Microtubule-determined plastid distribution during microsporogenesis in *Lilium longiflorum. J Cell Sci* 99: 21-31.

Tanaka I, Nakamura S (1989) Structural features of isolated generative cells and their protoplasts from pollen of some liliaceous plants. *Gamete Res.* 24: 361-374.

Traas JA, Doonan JH, Rawlins DJ, Shaw PJ, Watts J, Lloyd CW (1987) An actin network is present in the cytoplasm throughout the cell cycle of carrot cells and associates with the dividing nucleus. *J Cell Biol* 105: 387-395.

Upadhyaya MK, Nooden LD (1977) Mode of dinitroaniline herbicide action I. Analysis of the colchicine-like effects of dinitroaniline herbicides. *Plant Cell Physiol* 18: 1319-1330.

Upadhyaya MK, Nooden LD (1978) Relationship between the induction of swelling and the inhibition of elongation caused by oryzalin and colchicine in corn roots. *Plant Cell Physiol* 19: 133-138.

Vandre DD, Davis FM, Rao PN, Borisy GG (1984) Phosphoproteins are components of mitotic microtubule organizing centers. *Proc Natl Acad Sci USA* 81: 4439-4443.

Vantard M, Schellenbaum P, Fellous A, Lambert A-M (1991) Characterization of maize microtubule-associated proteins, one of which is immunologically related to *tau*. *Biochemistry* 30: 9334-9340.

Venverloo CJ, Libbenga KR (1987) Regulation of the plane of cell division in vacuolated cells I. The function of nuclear positioning and phragmosome formation. *J Plant Physiol* 131: 267-284.

Vera JC, Rivas CI, Maccioni RB (1989) Biochemical dissection of the role of the one-kilodalton carboxyl-terminal moiety of tubulin in its assembly into microtubules. *Biochemistry* 28: 333-339.

Verhoeven HA, Sree Ramulu K, Dijkhuis P (1990) A comparison of the effects of various spindle toxins on metaphase arrest and formation of micronuclei in cell-suspension cultures of *Nicotiana plumbaginifolia*. *Planta* 182: 408-414.

Voter WA, Erickson HP (1984) The kinetics of microtubule assembly: evidence for a two-stage nucleation mechanism. *J Biol Chem* 259: 10430-10438.

Wang D, Villasante A, Lewis SA, Cowan NJ (1986) The mammalian β–tubulin repertoire: Hematopoietic expression of a novel, heterologous β–tubulin isotype. *J Cell Biol* 103: 1903-1910.

Wani MC, Taylor HL, Wall ME, Coggon P, McPhail AT (1971) Plant antitumor agents. VI. The isolation and structure of taxol, a novel antileukemic and antitumor agent from *Taxus brevifolia*. *J Am Chem Soc* 93: 2325-2327.

Webster DR, Borisy GG (1989) Microtubules are acetylated in domains that turn over slowly. *J Cell Sci* 92: 57-65.

Wick SM (1985) Immunofluorescence microscopy of tubulin and microtubule arrays in plant cells. III. Transition between mitotic/cytokinetic and interphase microtubule arrays. *Cell Biol Int Rep* 9: 357-371.

Wick SM, Duniec J (1983) Immunofluorescence microscopy of tubulin and microtubule arrays in plant cells. Preprophase band development and concomitant appearance of nuclear envelope-associated tubulin. *J Cell Biol* 97: 235-243.

Wilms FHA, Wolters-Arts AMC, Derksen J (1990) Orientation of cellulose microfibrils in cortical cells of tobacco explants. Effects of microtubule-depolymerizing drugs. *Planta* 182: 1-8.

Zhang D, Wadsworth P, Hepler PK (1990) Microtubule dynamics in living dividing plant cells: Confocal imaging of microinjected fluorescent brain tubulin. *Proc Natl Acad Sci USA* 87: 8820-8824.

Zheng Y, Jung MK, Oakley BR (1991) γ–tubulin is present in *Drosophila melanogaster* and *Homo sapiens* and is associated with the centrosome. *Cell* 65: 817-823.

tissues and organs; the relationship of polyamines with other hormones (in relation to endogenous concentrations and ratios); possibly a different sensitivity of tissues to polyamines; and the conversion of free polyamines (active form) to conjugated polyamines.

Bearing in mind these considerations, the relationship between polyamines and several morphogenic processes and the possibility of controlling these by using polyamine biosynthetic inhibitors will be described in this review. The following aspects of plant morphogenesis will be analysed in relation to some *in vitro* model systems, namely: somatic embryogenesis, floral and vegetative bud formation, rhizogenesis, morphogenesis in transgenic tissues, and changes in cell size.

SOMATIC EMBRYOGENESIS

The complex developmental events leading to somatic embryogenesis require an ordered expression of morphogenic stimuli arising from endogenous genotype-specific factors, culture medium composition and physical environmental conditions. These developmental events include induction and activation of multipotent cells, resulting in the formation of embryogenic cell clumps, bipolar differentiation, embryo and plantlet development. The reproducible induction process and the methods described to select the different stages of somatic embryos (Giuliano *et al.*, 1983) have allowed much progress to be made in understanding the physiological and biochemical bases of somatic embryogenesis.

Much research has been undertaken to identify the biochemical markers that characterize the switch from unorganized to organized growth. In fact, unlike animal cells in which proliferation and differentiation are antithetical, in plant embryogenesis these two processes occur simultaneously. Among the biochemical markers considered, polyamines have a major function in embryo induction and development; they have been examined in relation to carrot somatic embryogenesis, which is, to date, the best model system. Montague *et al.* (1978) observed that the induction of embryogenesis by the removal of 2,4-dichlorophenoxyacetic acid (2,4-D) was accompanied by increased levels of the three major polyamines; also the activities of the enzymes for putrescine and spermidine synthesis generally increased under embryogenic conditions (Fienberg *et al.*, 1984).

Many studies demonstrating the essential role of polyamines have been made possible through the use of inhibitors of polyamine synthesis such as α-difluoromethylornithine (DFMO) and α-difluoromethylarginine (DFMA), specific irreversible inhibitors of ornithine- (ODC) and arginine decarboxylase (ADC) activity, respectively, which are responsible for putrescine synthesis. Feirer *et al.* (1984) found that DFMA but not DFMO provoked a nearly 50% reduction in embryo formation. Using a similar approach, Fobert and Webb (1988) reported that DFMO inhibited eggplant embryo formation, while causing a significant increase in dry weight. Methylglyoxal-bis(guanylhydrazone) (MGBG) and cyclohexylamine (CHA), competitive inhibitors of S-adenosylmethionine decarboxylase (SAMDC), involved in spermidine and spermine synthesis and of spermidine synthase, respectively, were found to reduce unorganized growth, embryogenesis and cellular polyamine levels in wild carrot cultures (Feirer *et al.*, 1985). Generally the effect of all these inhibitors was reversible by the addition of the corresponding polyamine to the medium. On the other hand, Bradley *et al.* (1984) reported that exogenous putrescine or arginine blocked the growth of carrot embryos beyond the globular stage, while in celery the addition of 0.1 mM spermidine to regeneration medium greatly enhanced plantlet formation from embryogenic clumps (Altman *et al.*, 1990).

Studies in our laboratory were aimed at identifying the metabolic features of cell commitment to embryogenesis (Mengoli *et al.*, 1989). The switch from cell proliferation to embryogenesis by removal of 2,4-D resides in the preembryogenic phase where, despite the presence of 2,4-D, some cell clumps, smaller and denser than others, organize "preglobular structures", as indicated by Choi and Sung (1989) and others. In this context, polyamine content, ADC and ODC activities as well as the effect of DFMO and canavanine, an unspecific competitive inhibitor of ADC activity, were examined prior to and during the formation of carrot somatic embryos. DFMO, added during the preembryogenic phase, lowered cellular putrescine levels without reducing growth, while surprisingly prolonging the commitment of the cells to embryogenesis (Fig. 1); on the contrary, canavanine completely blocked growth during the same phase. The activity of ODC, which was the predominant enzyme during the preembryogenic phase, was affected by DFMO in a cell-free system but not when supplied in the culture medium.

During the embryogenic phase treatment with canavanine for 36 h caused an almost complete inhibition of any further development, while DFMO-treated globular or heart-

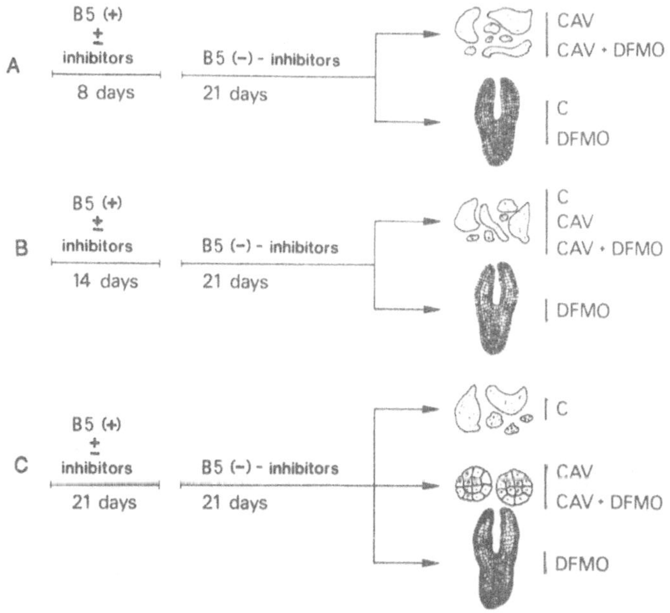

Figure 1. Scheme of the effect on somatic embryogenesis of the transfer of cultured carrot cells from preembryogenic medium [B5(+) with 2,4-D] supplemented with 1 mM canavanine (CAV) and/or 1 mM difluoromethylornithine (DFMO) for different times (A,B,C) to embryogenic medium [B5(-) without 2,4-D] lacking inhibitors. C=untreated controls.

shaped embryos developed normally, were larger and showed a higher conversion rate to plantlets than untreated controls. A concomitant increase in polyamine levels and increments in fresh weight and protein content were observed after treatment with DFMO during the embryo development phase. As a general trend, in control cultures spermidine and spermine contents, expressed on a per unit basis, increased during the progression from globular to heart and torpedo stage; putrescine was the main polyamine in the heart stage, whereas spermidine was more abundant in torpedo stage.

A better understanding of the factors regulating somatic embryogenesis is of interest for both basic and applied research. In the attempt to increase the efficiency of plantlet formation from somatic embryos in *Vitis vinifera*, a comparative study was carried out in our laboratory on free polyamine levels, ODC and ADC activities in somatic and zygotic embryos (Faure *et al.*, 1991). While previous work has dealt with polyamine pattern during the development and germination of zygotic embryos (Bagni, 1970; Sen *et al.*, 1981), only a few comparative studies have been carried out on zygotic and somatic embryos of the same species, e.g. *Mangifera indica* (Litz and Schaffer, 1987). Grape somatic embryos, although produced at a high frequency, are mostly teratologic and display abnormal growth and cell proliferation (Faure, 1990; Fig. 2); this causes a very low conversion rate into plantlets (<5%). Compared to zygotic embryos, somatic embryos showed higher levels of free polyamines and a high putrescine/ spermidine ratio. On a per unit basis, an accumulation of free polyamines occurred in the late torpedo stage, which coincided with the beginning of abnormal cell proliferation. Concomitantly, a very high ADC and an even higher ODC activity occurred. These data suggest that the low yield in plantlets from somatic embryos could be due to the high polyamine content and/or to an inadequate putrescine/spermidine

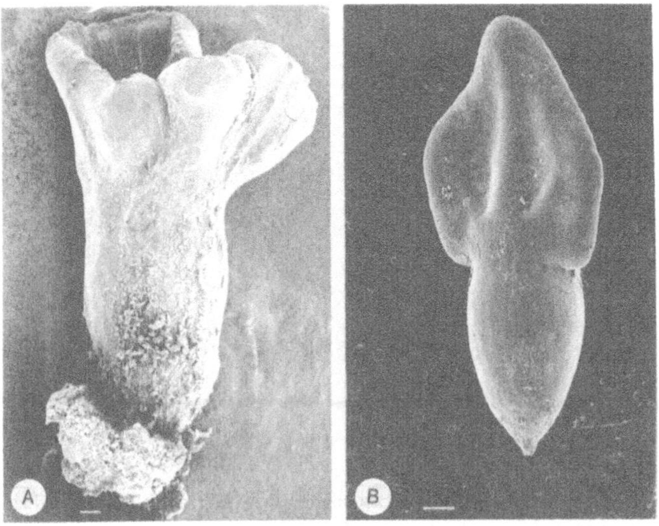

Figure 2. Scanning electron micrographs of late torpedo stage somatic embryo (A) and torpedo stage zygotic embryo (B) of *Vitis vinifera* cv. Grenache noir. Bar = 100 μm.

ratio. Indeed, a significant decrease in this ratio was also observed by Altman *et al.* (1990) during the normal progression from globular stage to plantlet in celery somatic embryogenesis. Moreover, it could be that the mineral medium is not suitable for grape somatic embryos which might consequently be stressed and react by accumulating putrescine, leading to excessive cell proliferation.

In conclusion, although many data suggest a causal involvement of polyamines in somatic embryogenesis, most of the observed changes in polyamine levels may not be related specifically to this process, since similar changes were observed during the progression of other morphogenic programs (Torrigiani *et al.*, 1987, 1989).

FLORAL AND VEGETATIVE BUD FORMATION

Polyamines are present inside the cell, at physiological pH values, partly in an ionized form which is responsible for the interactions with negatively charged compounds, such as nucleic acids, phospholipids, etc. (Bachrach, 1989). They are also partly covalently linked to low or high molecular weight compounds; conjugated polyamines are defined by determination after acidic hydrolysis of the supernatant or of the perchloric acid-/trichloracetic acid (TCA)-precipitable material, respectively. In many cases the structure of the polyamine partner molecules has not been determined. In tobacco and other plants polyamines bind to hydroxycinnamic acids (Flores and Martin-Tanguy, 1991) to form amides which are the most common form of TCA-soluble polyamine conjugates in higher plants. Protein-bound polyamines which are well known in animal cells, have been described in a few species, such as tobacco, *Helianthus tuberosus, Medicago sativa*, oat and *Petunia* (Apelbaum *et al.*, 1988; Margosiak *et al.*, 1990; Mizrahi *et al.*, 1989; Serafini-Fracassini *et al.*, 1988). Until now no acetylated polyamines, the major polyamine metabolites in animal cells, have been detected in higher plants.

Tobacco thin layers, the plant model system introduced by Tran Thanh Van (1973), represent a very useful tool in studying patterns of free and conjugated polyamines during specific morphogenic programs. It is known that the hormone balance affects polyamine biosynthesis (Bagni, 1989; Tiburcio *et al.*, 1990), therefore a comparative study of polyamine titers during *in vitro* floral and vegetative bud formation in *Nicotiana tabacum* cv. Samsun, has helped in the understanding of the possibly different involvement of free and conjugated polyamines in the two morphogenic processes (Torrigiani *et al.*, 1987).

The model configuration used by Tran Thanh Van *et al.* (1974) was based on the possibility of switching the organogenic program only by changing the auxin/cytokinin ratio. In this system, exogenous spermidine induced flower initiation in cultures which would otherwise have formed vegetative buds (Kaur-Sawhney *et al.*, 1988). When the same hormonal ratio (1 µM auxin/1 µM cytokinin) was adopted to induce vegetative or floral programs but from a different site of excision (stem and pedicel internode explants, respectively, from plants in the unripe fruit stage), the first flower buds appeared on day 8 in culture, while the first vegetative buds appeared on day 12. Thus, the completion of the vegetative program was shifted forward by 3-4 days with respect to the floral one. The monitoring of free and conjugated polyamines throughout the culture period yielded a similar pattern in the two programs.

polyamines in tobacco cultures indicate that the availability of free polyamines, also necessary for conjugate synthesis, is required in the neoformation of buds. It is now also evident that both free and conjugated polyamines are involved *in vitro* in the initiation and completion of the reproductive as well as the vegetative program.

RHIZOGENESIS

Understanding the factors controlling adventitious root formation is an important contribution to basic and applied research. With regard to the latter, and particularly in agriculture and forestry, there is a need to optimize the rooting ability of cuttings and micropropagated shoots. With this in mind, a study was conducted by us on the role of polyamines and ethylene in adventitious root formation using a well-established micropropagation protocol for *Prunus avium* (Biondi *et al.*, 1990).

Correlations between free polyamine accumulation and root formation have been observed in *Passiflora* and *Datura innoxia* leaf explants (Chriqui *et al.*, 1986; Desai and Mehta, 1985) and in mung bean hypocotyl cuttings (Friedman *et al.*, 1982). In the latter work, exogenously applied polyamines did not promote rooting, whereas Jarvis *et al.* (1983) reported an enhancement in the mean number of roots developing on stem cuttings of the same species upon treatment with spermine. In both studies concerning mung bean, various inhibitors of polyamine biosynthesis inhibited rooting.

While auxins seem to be the most effective substances for the promotion of adventitious root formation, we are far from understanding their specific function and the way in which they interact with other growth factors. It is known, for instance, that there is a close relationship between auxin and ethylene, but data regarding the direct action of ethylene on rhizogenesis are still controversial (Jarvis, 1986).

It is now well accepted that in higher plants methionine, after its conversion to S-adenosylmethionine (SAM), forms 1-aminocyclopropane-1-carboxylic acid (ACC) and hence ethylene (Yang and Hoffman, 1984). In the polyamine biosynthetic pathway, SAM is decarboxylated and serves as donor of the aminopropyl moiety which converts putrescine to spermidine and spermidine to spermine. Thus, SAM plays a pivotal role in the biosynthesis of ethylene and polyamines and the regulation of its metabolic fate may have important physiological implications also in rooting.

Proliferating shoots of an easy-to-root clone of *Prunus avium* were established from axillary buds. During the rooting phase on indole-3-butyric acid (IBA)-containing medium, data on the endogenous content of polyamines measured separately in the apical and basal portions of shoots and in stems and leaves revealed that putrescine, the most abundant polyamine, and to some extent spermidine, also present in all shoot portions, peaked between days 9 and 11. Root emergence begins approximately on days 12-14 and thus, it would seem that polyamines accumulate during the organization of root primordia. However, in a less complex model system, such as tobacco thin layers, polyamine accumulation begins during the early stages of meristemoid formation (see below).

Exogenously supplied putrescine or spermidine, in the presence of optimal or suboptimal levels of IBA, had no effect on the percentage of rooted shoots (rooting percentage) or on the mean number of roots per shoot (root density), except 500 μM spermine which was strongly inhibitory. This treatment also provoked a marked alteration in endogenous polyamine titers, namely a large accumulation of putrescine and spermine. The latter may have exerted a feedback inhibition of putrescine conversion to spermidine.

DFMO, alone or in combination with DFMA, and CHA plus MGBG, in the 0.1 - 5 mM range, all inhibited rooting percentage and root density. This inhibition was only partially reversed by the addition of the corresponding polyamine (putrescine or spermidine). Cellular polyamine levels were significantly reduced by DFMO and DFMA, but not by CHA and MGBG.

Incorporation of [3,4-^{14}C]methionine into ethylene was drastically reduced by the ethylene synthesis inhibitor aminoethoxyvinylglycine (AVG) and increased with increasing concentrations of IBA. AVG also led to an increase in labeled putrescine and methionine incorporation into spermidine. This lends support to the fore-mentioned interdependence between the biosynthetic pathways of ethylene and polyamines. As to its effect on rooting, AVG enhanced the mean number of roots per shoot while delaying their elongation; ACC reduced rooting percentage.

Thus, in this system polyamines may play a significant role at least in some stages of adventitious root formation. Endogenous polyamine levels are sufficient to support this process; hence, an exogenous supply is unnecessary and their depletion is detrimental. Moreover, the polyamine and ethylene biosynthetic pathways are, at least to some extent, competitive. Finally, although IBA promotes rooting as well as ethylene synthesis, AVG, which drastically reduces the latter, also promotes root formation. Thus, the auxin effect on root formation cannot be directly related to its ability to enhance ethylene synthesis.

Rhizogenesis was also studied in the tobacco thin layer model system (Altamura *et al.*, 1991b; Torrigiani *et al.*, 1989). On explants, excised from the stem of *Nicotiana tabacum* cv. Samsun plants in the complete green fruit stage, roots appeared on day 12 in culture, as did vegetative buds. When the program was completed, about 80-90% of explants exhibited 7-8 roots/explant. Both free putrescine and spermidine rose early in culture until day 10 and then decreased (Fig. 3), while conjugated putrescine showed peaks of accumulation around day 15 (Figs 4 and 5). These patterns strictly resemble those previously described for floral and especially vegetative bud formation in the order of magnitude of the different classes, in their reciprocal ratios and in the correspondence between polyamine accumulation and the progression of meristemoid organization. These correlative data were corroborated in the same model system by the fact that the inhibition of polyamine biosynthesis and accumulation resulted in a corresponding inhibition of the rhizogenic response (Fig. 6). The inhibitors of putrescine synthesis reduced strongly (DFMO) or almost totally (DFMA+ DFMO) free and conjugated putrescine levels and rhizogenesis. Also when supplied separately, DFMO and DFMA caused inhibition of free and conjugated polyamine accumulation and rhizogenesis in tobacco leaf explants (Burtin *et al.*, 1990). CHA scarcely affected rhizogenesis although spermidine levels were strongly depleted, while putrescine rose at the end of the culture period. MGBG almost totally inhibited the rhizogenic response and reduced spermidine levels on days 15-18, but caused an unexpected spermidine and putrescine accumulation on day 30. Our results suggest that both enzymes of putrescine synthesis could be involved in rhizogenesis although we do not know which enzyme activity is involved in

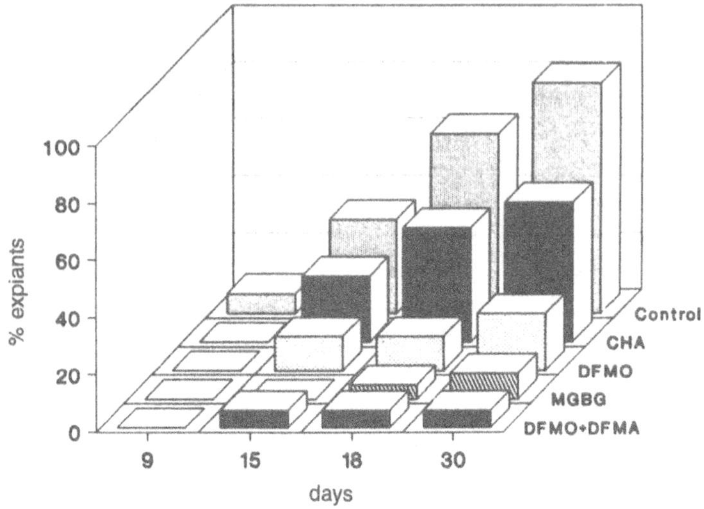

Figure 6. Percentage at different times in culture of rooting tobacco thin layer explants in the presence or absence of 1 mM DFMO, DFMA or MGBG or 5 mM CHA.

which physiological event (cell division or cell enlargement). Other model systems, such as *Hyoscyamus muticus* hairy roots and cultured tobacco cells have provided some indications that cell enlargement is enhanced by blocking ODC activity with DFMO (Berlin and Forche, 1981 and section below).

The inhibition of spermidine synthesis by CHA was not reflected in an equally dramatic effect on the rhizogenic response. On the contrary, the MGBG-induced response was different in that it inhibited both the spermidine increase and rhizogenesis. By supplying the reversing polyamine together with the drug at the onset of culture, there was no recovery of root formation (Altamura *et al.*, 1991b), while a 50% recovery was achieved when spermidine was added in MGBG-treated explants at the time of its maximum depletion and at a concentration comparable to the endogenous one (0.3 - 0.4 mM); in CHA-treated explants, in which rhizogenesis was not strongly affected, it did not cause any reversion (unpublished results).

Thus there are strong indications that free polyamines are required for the rhizogenic process to occur and that conjugated polyamines are also involved (see also Tiburcio *et al.*, 1989); this has also been found for the other morphogenic programs.

It remains to be clarified whether the involvement of conjugated polyamines is a particular feature of tobacco or of the *Solanaceae* and, assuming that TCA-soluble conjugates are hydroxycinnamoylamides, what the nature of the TCA-precipitable partner molecules of polyamines is, in which cell compartment they are mainly located and what the mechanism is by which they might affect growth and development.

MORPHOGENESIS IN TRANSGENIC TISSUES

Agrobacterium rhizogenes is a naturally-occurring soil bacterium which infects almost exclusively dicotyledonous plants and induces the formation of adventitious roots at the site of infection (Tepfer, 1984). These roots are prolific, highly-branched and are known as hairy roots. They can be isolated and, after treatment with antibiotics, grown in axenic culture in the absence of hormones. Both *in vivo* and *in vitro* the hairy root response derives from the insertion and expression in the host cell of a specific DNA sequence called transferred-DNA (T-DNA) on the root-inducing (Ri) plasmid belonging to the bacterium. The T-DNA of the *A. rhizogenes* strains examined to date can be subdivided into a right (T_R) and left (T_L) portion which can be independently transferred. The T_L-DNA carries the *rol* genes known to be responsible for the induction of the rhizogenic response.

Rapidly growing hairy root cultures with their prolific formation of lateral roots seem to be a good model to elucidate the involvement of polyamines in root growth and branching. A study was conducted on the effect of the most commonly used inhibitors of polyamine biosynthesis on root growth and morphology as well as on their endogenous titers in hairy roots of *Hyoscyamus muticus.*

DFMO, CHA or MGBG affected root growth and morphology by reducing the number of lateral roots formed by day 16 (Fig. 7). At the same time, the inhibitory effect of CHA and in particular of MGBG on primary root growth was evident. DFMO, on the other hand, stimulated elongation growth of primary roots (57% above untreated controls) and of lateral roots. When DFMO was supplied together with DFMA, primary root length was not significantly affected with respect to untreated cultures, while lateral root growth was inhibited. In addition, DFMO, alone or in combination with DFMA, exerted a very marked reduction in root hair formation on neoformed lateral roots (Fig. 8). For comparative purposes some preliminary observations were made on untransformed root cultures of the same species which were able to grow under the same culture conditions. Untransformed roots, however, have a much reduced branching capacity and the few lateral roots formed are

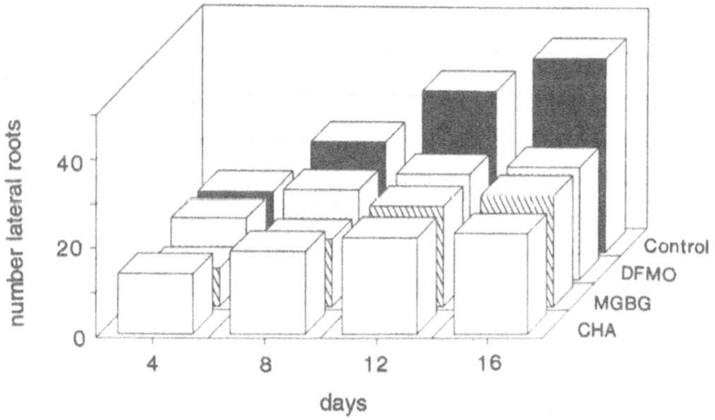

Figure 7. Effect of 1 mM DFMO or MGBG and of 5 mM CHA on the number of lateral roots formed on cultured hairy root explants of *Hyoscyamus muticus.*

roots coincided with lower ADC and ODC activities; in addition, a high amount of putrescine bound to TCA-precipitable molecules was detected early on in culture, when T roots displayed the fastest growth rate. Histological observations indicating that cell size was substantially the same in the three lines suggested that growth occurred primarily by cell division. Assuming that the partner molecules of this bound putrescine are mainly proteins, one could hypothesize that it plays a regulatory role in cell division as suggested by Serafini-Fracassini *et al.* (1989) in relation to the cell cycle in *Helianthus tuberosus* tuber explants. In conclusion, at least in this model system, polyamine levels and related enzyme activities seem to be linked to differences in growth kinetics (rate of cell division and root elongation) and to the presence of numerous lateral root meristems, rather than being a direct consequence of foreign gene expression.

Transgenic tobacco plants contain lower levels of polyamines than normal ones. Martin-Tanguy *et al.* (1990) suggested that genes carried by the Ri T-DNA either intervene directly in polyamine metabolism or that polyamine metabolism is linked to direct effects of Ri T-DNA expression. Further investigations were carried out by Chriqui *et al.* (1990) showing lower free auxin levels in shoot apices of T and T' tobacco plants compared to untransformed ones. In our laboratory (Mengoli *et al.*, 1992a, b), a more detailed study of the physiological bases underlying the hairy root phenotype indicated the following: 1) this phenotype is differently modulated under *in vivo* (greenhouse) versus *in vitro* culture conditions as observed also by Ooms *et al.* (1985) in *Brassica napus*, possibly due to a stronger selection pressure *in vivo* than *in vitro*; 2) the comparison of protein, amino acid and polyamine levels between normal and transformed plants revealed alterations in nitrogen metabolism, with apices and leaves of transformed plants displaying significantly lower levels of glutamine and asparagine; 3) under greenhouse conditions, the transformed tissues always exhibited lower levels of conjugated polyamines than the normal ones, with TCA-soluble bound polyamines undetectable in shoot apices and leaves of both transformed and untransformed *in vitro* plantlets; 4) concerning changes in the polyamine metabolic pathways, diamine oxidase activity was higher in T and even more in T' *in vitro* plants than in controls and this was particularly evident in shoot apices. A correlation was found between endogenous polyamine levels and their biosynthetic enzyme activities (ADC, ODC) in apices but not in leaves of *in vitro* plants, possibly due to a different compartmentation of enzymes, substrates and products.

CHANGES IN CELL SIZE

Root histogenesis in tobacco thin layers was extensively studied (Altamura *et al.*, 1991a), as well as the effect of various inhibitors of polyamine biosynthesis on root formation (see section on Rhizogenesis). Subsequently, the cytological modifications induced by treatment with these inhibitors was examined in the course of the same morphogenic program.

Cell dimensions (length and width in radial longitudinal section), measured in each cell layer at different times in culture, revealed the occurrence of two main cytological responses: cell hypertrophy, in which both length and height were significantly greater than

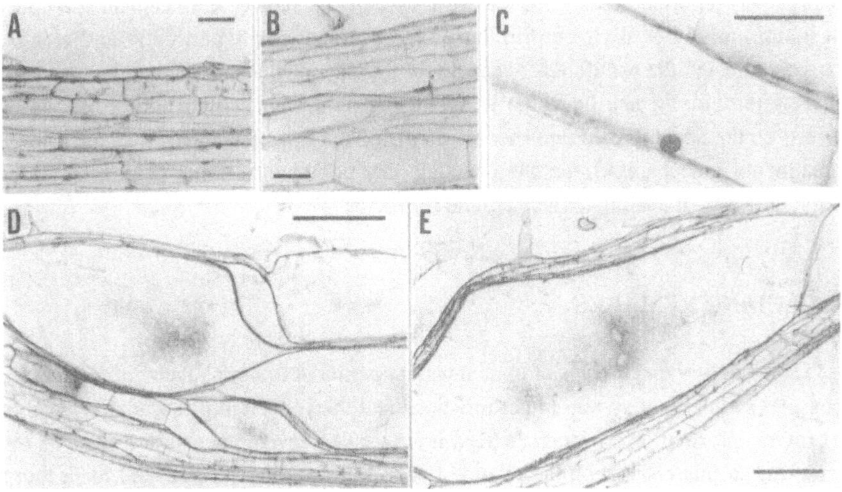

Figure 9. (A) Cell hypertrophy and/or elongation induced by the inhibition of polyamine biosynthesis in tobacco thin layer system (stains according to Altamura *et al.*, 1991a). Explant at day 0 (L.S., bar = 60 μm); (B): Hypertrophy in parenchymal layer in contact with the medium (DFMO + DFMA treatment, day 9, bar = 60 μm, L.S.); (C): Elongated cell in the parenchymal layer in contact with the medium (DFMO treatment, day 9, bar = 20 μm, L.S.); (D): Hypertrophy in the subepidermis and median parenchyma (CHA treatment, day 18, bar = 200 μm, L.S.). (E) Hypertrophic cell of the median parenchyma (CHA treatment, day 18, bar = 200 μm, L.S.).

in untreated controls and cell elongation in which length was affected much more than height. Elongated cells were observed in controls and DFMO did not significantly enhance this cytological response (Fig. 9C). On the other hand, DFMO+DFMA, MGBG and CHA also provoked the appearance of a few elongated cells, but the most dramatic effect was the occurrence of many hypertrophic cells (Fig. 9B, D, E) which were never observed in controls. This phenomenon was strongly reduced when the reversing polyamine was supplied with the inhibitor (MGBG + spermidine, CHA + spermidine, DFMO + DFMA + putrescine).

The target layer of cell hypertrophy and/or cell elongation was the parenchyma layer in contact with the medium (Fig. 9B, C). The maximum increase in cell size, expressed as mean cell area and referred to controls was 2-fold in DFMO + DFMA (day 9) and MGBG (day 15) and 5-fold in CHA (day 18), and about 4-fold (DFMO + DFMA, MGBG) and 7-fold (CHA) with respect to uncultured explants (day 0). Only in the case of CHA treatment, both the cells of the subepidermis and of the median parenchyma showed hypertrophy (Fig. 9D and E). The supepidermal and the underlying parenchyma cells are the first to undergo cell division and organize root meristemoids (Altamura *et al.*, 1991a) and probably for this reason, they are particularly sensitive to the drugs. The reversion of cell hypertrophy and elongation by polyamines mainly occurred at the day in culture in which the inhibitor alone had induced the greatest expression of the phenomenon. On the same days, maximum depletion of endogenous polyamines was also attained (Altamura *et al.*, 1991b). Berlin and

Forche (1981) reported that DFMO-treated tobacco cells showed enlargement and concomitant inhibition of cell division. Similarly we observed that hypertrophy and elongation negatively affected the proliferative capacity of thin layers, hence reducing the formation of root meristemoids. Since no negative influence of the polyamine inhibitors was ever observed on the development into root primordia of the few meristemoids formed (fewer in all treatments except CHA), we can conclude that polyamines affect mainly the phase of proliferative growth leading to meristemoid formation.

CONCLUDING REMARKS

There are two major ways of obtaining data pertinent to morphogenesis: studies on the response(s) of the plant system to perturbations and the collection of correlative data. Such data provide information on what is *essential* for a given developmental process. The results summarized in this chapter on the relation between polyamines and several plant morphogenic processes have demonstrated that a considerable amount of correlative evidence is now available. Any given essential factor may, however, be one of very many and thus the key to understanding a complex phenomenon such as morphogenesis lies in the relationships between them.

In fact the involvement of polyamines in the regulation of protein (Igarashi *et al.*, 1977) and RNA synthesis, of DNA conformational changes (Feuerstein and Marton, 1991) and in the induction and control of cell division (Bagni, 1989), all crucial events of cell growth, leads us to believe that a specific signal(s) or event(s) triggered by polyamines (and not by other growth substances or secondary messengers) may indeed be possible.

At the present time the other useful approach towards a better understanding of the role of polyamines in morphogenesis resides in the use of specific inhibitors of polyamine biosynthesis, namely DFMO, DFMA, MGBG and CHA which, however, could have undesired side effects at high concentrations. In our opinion, the main difficulty in these studies arises from the fact that cell division, cell expansion and morphogenesis in plants are all tightly linked.

The use of such inhibitors in simple model systems like spore germination and protonemal growth in *Adiantum capillus-veneris* (Galston *et al.*, 1990) as well as cultured tobacco cells (Berlin and Forche, 1981), thin layer systems (Altamura *et al.*, 1991b) and somatic embryogenesis (Mengoli *et al.*, 1989) suggests that polyamines are essential for cell division. In fact, by perturbing polyamine metabolism with biosynthetic inhibitors, cell division and in some cases cell elongation can be blocked. The response depends on the type of inhibitor and/or on the particular plant system used. If this scheme is correct, the main role of polyamines in morphogenic processes resides in the initiation of root and shoot primordia and in the early phases of embryo formation. Thus the real possibility exists that in most, if not all, the morphogenic processes examined, polyamines are involved only during the phase of cell division and have little to do with morphogenesis *per se*. It would be useful to have a well-characterized model system in which to intervene after the cell division phase, that is at the time of cell elongation which leads to organ formation and emergence, in order to better understand the real involvement of polyamines in these processes.

Sometimes inhibition of polyamine biosynthesis can produce bizarre morphogenic consequences like that induced in *Adiantum* protonemata (Galston *et al.*, 1990), tobacco cells (Berlin and Forche, 1980), *Hyoscyamus muticus* hairy roots, and carrot somatic embryos (Mengoli *et al.*, 1989). In this context it would be instructive to study the effect of these inhibitors on cell wall properties, on the cytoskeleton and on tip growth.

However, we cannot exclude other considerations, common to other plant growth regulators, such as pleiotropic effects of polyamines, but especially the ratio between them and other plant hormones; this can be critical in inducing or changing the morphogenic program. The classical experiment conducted by Skoog and Schmitz (1972) using the tobacco pith system aimed at demonstrating the role of cytokinins in the presence or absence of auxins could be useful as a basic approach to clarify the morphogenic effects of polyamines. In fact, many data suggest that endogenous polyamine titers change in response to exogenous plant hormones (namely auxins, cytokinins and gibberellins), but in some cases endogenous plant hormones change also in response to added polyamines (Bagni, 1989).

While there are some indications that polyamines act as mediators of auxin action (Cocucci and Bagni, 1968; Galston, 1983), more recent work on the involvement of polyamines in gibberellin-promoted internodal growth again suggest that they are needed for the cell division phase preceding gibberellin-promoted internodal cell elongation (Smith *et al.*, 1985).

Recent evidence demonstrating that polyamine uptake is enhanced by auxin in the presence of Ca^{2+} (Kanchanapoom *et al.*, 1991) suggests the existence of an auxin-polyamine relationship not only in the regulation of polyamine biosynthesis and oxidation (Bagni, 1989; Srivastava *et al.*, 1977), but also in the regulation of their transport. The flux of polyamines can be important in many organogenic processes, a hypothesis advanced for the early events of flowering in *Xanthium strumarium* (Galston *et al.*, 1990). The movement of polyamines and/or of their metabolites from expanding leaves to buds induced to flower was recently studied in soybean (Caffaro *et al.*, 1992); these data are consistent with studies on polyamine translocability (Bagni and Pistocchi, 1991) and follow the general rules of transport that apply to flower induction and evocation (Vince-Prue, 1975). Although the numerous unidentified products derived from labeled polyamines supplied to the leaf suggest the need for further studies, true polyamine synthesis and accumulation seem to be under phytochrome control (Caffaro *et al.*, 1992; Galston *et al.*, 1990).

The possible control of flowering by polyamines or their hydroxycinnamoyl conjugates must as yet remain speculative. In fact, hydroxycinnamoyl-conjugated polyamines have other implications in addition to their role in flowering. The possibility exists that polyamine conjugation to a hydroxycinnamoyl moiety is important in the regulation of free polyamine titers and/or in setting aside these phenolic compounds known to induce inhibitory growth effects, from cell metabolism. In fact cinnamic acids inhibit seed germination and growth; hordatines, which are dimers of coumaroylagmatine, have antifungal properties in barley seedlings (Stoessl and Unwin, 1970). Further, 2-hydroxyputrescine amides of ferulic acid and p-coumaric acid are formed in wheat in response to rust infection (Samborski and Rohringer, 1970) and in tobacco in response to mosaic virus infection (Martin-Tanguy, 1985). In addition, cinnamic acids such as p-coumaric and ferulic acids are known to be the immediate precursors of lignin (Grisebach, 1978) and together with caffeic acid are utilized for the synthesis of some flavonoids like umbelliferone, aesculine and scopoletine. Thus, the multifunctional aspects and the different metabolic fates of these compounds leave every possibility open as to their direct role in growth regulation.

Polyamines conjugated to TCA-precipitable molecules, of which proteins represent a significant part, have no doubt an important role to play in growth and development. Very little is known about the exact nature of the partner molecules, although work is in progress in order to identify them and to characterize the transglutaminase-like activity which binds polyamines to proteins (Serafini-Fracassini *et al.*, 1989). Thus one could postulate a regulatory role of polyamines mediated by post-translational modifications of structural and/or enzymatic proteins. In this context an interesting paper by Pfosser *et al.* (1990) reported increases in free and TCA-precipitable putrescine closely correlating with specific phases, i.e. preceding S phase, of the synchronous cell cycle in tobacco cell cultures, as previously shown to occur for free polyamines in *Helianthus tuberosus* tuber explants (Serafini-Fracassini *et al.*, 1980).

Embryogenesis seems to be one of the most exciting areas to study the relationships with polyamines due to the fact that only the inhibitors of ADC and not those of ODC are able to block embryo formation. This suggests a dual role of these enzymes in morphogenic processes. Finally, the use of transgenic plants could provide in future new information on the function of polyamines in regulating plant morphogenesis. At the present time there is no information in this direction even though Hamill *et al.* (1990) have introduced the yeast gene coding for ODC into *Nicotiana rustica* as a means of perturbing alkaloid synthesis. The use of transgenic plants regenerated from roots transformed with *Agrobacterium rhizogenes* seems to suggest a deregulation of morphogenesis. Whether the low levels of polyamines associated with transgenic plants can be related directly to the expression of one or more genes involved in polyamine metabolism or whether they are a secondary event deriving from changes in the content of and/or sensitivity to other hormones (auxins and cytokinins) is far from being understood because the molecular genetic analysis of plant polyamine synthesis is only beginning (Bell and Malmberg, 1990).

The possibility of cloning the major genes involved in polyamine biosynthesis and the isolation of mutants lacking one or more pathways of their biosynthesis will, in the near future, open the way to the manipulation of endogenous polyamine levels. This also will provide information on the different expression of these genes, their promotors and tissue-specificity. Knowledge of all these factors is absolutely required to understand the specific role of polyamines in growth and development.

ACKNOWLEDGEMENTS

Research supported by National Research Council of Italy, Special Project RAISA, Sub-project No. 2, Paper No. 381.

REFERENCES

Altamura MM (1989) Origin and development of flowers and inflorescences in tobacco thin cell layers. *Giorn Bot Ital* 123:23-28.

Altamura MM, Capitani F, Serafini-Fracassini D, Torrigiani P, Falasca G (1991a) Root histogenesis from tobacco thin cell layers. *Protoplasma* 161:31-42.

Altamura MM, Torrigiani P, Capitani F, Scaramagli S, Bagni N (1991b) *De novo* root formation in tobacco thin layers is affected by inhibition of polyamine biosynthesis. *J Exp Bot* 42:1575-1582.

Altman A, (1989) Polyamines and plant hormones. *In*: The Physiology of Polyamines, Bachrach U, Heimer YM (eds), vol 2, CRC Press, Boca Raton, Florida, pp 121-145.

Altman A, Nadel BL, Falash Z, Levin N (1990) Somatic embryogenesis in celery: induction, control and changes in polyamines and proteins. *In*: Progress in Plant Cellular and Molecular Biology, Nijkamp HJJ, van der Plas LHW, van Aartrijk J (eds), Kluwer Academic Publishers, Dordrecht, pp 454-459.

Apelbaum A, Canellakis ZN, Applewhite PN, Kaur-Sawhney R, Galston AW (1988) Binding of spermidine to a unique protein in thin layer tobacco tissue culture. *Plant Physiol* 88:996-998.

Bachrach U, Heimer YM (1989) The Physiology of Polyamines, vol 2, CRC Press, Boca Raton, Florida.

Bagni N (1966) Aliphatic amines and a growth-factor of coconut milk as stimulating cellular proliferation of *Helianthus tuberosus* (Jerusalem artichoke) *in vitro*. *Experientia* 22: 732-733.

Bagni N (1970) Metabolic changes of polyamines during the germination of *Phaseolus vulgaris*. *New Phytol* 69: 159-164.

Bagni N (1989) Polyamines in plant growth and development. *In*: The Physiology of Polyamines, Bachrach U, Heimer YM (eds), vol 2, CRC Press, Boca Raton, Florida, pp 107-120.

Bagni N, Pistocchi, R (1991), Uptake and transport of polyamines and inhibitors of polyamine metabolism in plants. *In* : Biochemistry and Physiology of Polyamines in Plants, Slocum RD, Flores HE (eds), CRC Press, Boca Raton, Florida, pp 105-120.

Bell E, Malmberg RL (1990) Analysis of a cDNA encoding arginine decarboxylase from oat reveals similarity to the *Escherichia coli* arginine decarboxylase and evidence of protein processing. *Mol Gen Genet* 227:431-435.

Berlin J, Forche E (1981) DL-difluoromethylornithine causes enlargement of cultured tobacco cells. *Z Pflanzenphysiol* 101:277-282.

Biondi S, Bagni N, Sansovini A (1986) Dicyclohexylamine uptake and effects on polyamine content in cultured cotyledons of radiata pine. *Physiol Plant* 66:41-45.

Biondi S, Diaz T, Iglesias I, Gamberini G, Bagni N (1990) Polyamines and ethylene in relation to adventitious root formation in *Prunus avium* shoot cultures. *Physiol Plant* 78:474-483.

Bradley PM, El-Fiki F, Giles KL (1984) Polyamines and arginine affect somatic embryogenesis of *Daucus carota*. *Plant Sci Lett* 34:397-401.

Burtin D, Martin-Tanguy J, Paynot M, Carr M, Rossin N (1990) Polyamines, hydroxycinnamoyl-putrescines, and root formation in leaf explants of tobacco cultivated *in vitro*. *Plant Physiol* 93:1398-1404.

Caffaro S, Scaramagli S, Antognoni F, Bagni N (1992) Polyamine content and translocation in soybean plants. *J Plant Physiol*, in press.

Choi JH, Sung ZR (1989) Induction, commitment and progression of plant embryogenesis. *In* : Plant Biotechnology, Kung SD, Arntzen CJ (eds), Butterworths, Boston, pp 141-145.

Chriqui D, D'Orazi D, Bagni N (1986) Ornithine and arginine decarboxylases and polyamine involvement during *in vivo* differentiation and *in vitro* dedifferentiation of *Datura innoxia* leaf explants. *Physiol Plant* 68:589-596.

Chriqui D, Tepfer M, Prinsen E, van Onkelen H (1990) Composantes morphogénétiques et hormonales du phénotype hairy root chez *Nicotiana tabacum* cv Xanthi. *Bull Soc Bot Fr, Actual bot* 137:79-91.

Cocucci S, Bagni N (1968) Polyamine induced activation of protein synthesis in ribosomal preparation from *Helianthus tuberosus* tissue. *Life Sci* 7:113-120.

Cohen E, Malis Arad S, Heimer YM, Mizrahi Y (1982) Participation of ornithine decarboxylase in the early stages of tomato fruit development. *Plant Physiol* 70:540-543.

Desai HV, Mehta AR (1985) Changes in polyamine levels during shoot formation, root formation, and callus induction in cultured *Passiflora* leaf discs. *J Plant Physiol* 119:45-53.

Dumortier FM, Flores HE, Shekhawat NS, Galston AW (1983) Gradients of polyamines and their biosynthetic enzymes in coleoptiles and roots of corn. *Plant Physiol* 72:915-918.

Evans PT, Malmberg RL (1989) Do polyamines have roles in plant development?. *Annu Rev Plant Physiol Plant Mol Biol* 40: 235-269.

Fahn A (1967) Plant Anatomy. Pergamon Press, Oxford.

Faure O (1990) Embryons somatiques de *Vitis rupestris* et embryons zygotiques de *Vitis* sp: morphologie, histologie, histochimie et développement. *Can J Bot* 68:2305-2315.

Faure O, Mengoli M, Nougarède A, Bagni N (1991) Polyamine pattern and biosynthesis in zygotic and somatic embryo stages of *Vitis vinifera*. *J Plant Physiol* 138:545-549.

Feirer RP, Mignon G, Litvay JD (1984) Arginine decarboxylase and polyamines required for embryogenesis in the wild carrot. *Science* 223:1433-1435.

Feirer RP, Wann SR, Einspahr DW (1985) The effects of spermidine synthesis inhibitors on *in vitro* plant development. *Plant Growth Regul* 3:319-327.

Feuerstein BG, Marton LJ (1991) Specificity and binding in polyamine/nucleic acid interaction. *In* : The Physiology of Polyamines, Bachrach U, Heimer YM (eds), vol 1, CRC Press, Boca Raton, Florida, pp 109-124.

Fienberg AA, Choi JH, Lubich WP, Sung ZR (1984) Developmental regulation of polyamine metabolism in growth and differentiation of carrot culture. *Planta* 162:532-539.

Flores HE, Young ND, Galston AW (1985) Polyamine metabolism and plant stress. *In*: Cellular and Molecular Biology of Plant Stress, Key JL, Kosuge T (eds), Alan R Liss, New York, pp 93-114.

Flores HE, Martin-Tanguy J (1991) Polyamines and plant secondary metabolites. *In:* Biochemistry and Physiology of Polyamines in Plants, Slocum RD, Flores HE (eds), CRC Press, Boca Raton, Florida, pp 57-76.

Fobert PR, Webb DT (1988) Effects of polyamines, polyamine precursors, and polyamine biosynthetic inhibitors on somatic embryogenesis from eggplant (*Solanum melongena*) cotyledons. *Can J Bot* 66:1734-1742.

Friedman R, Altman A, Bachrach U (1982) Polyamines and root formation in mung bean hypocotyl cuttings. I. Effects of exogenous compounds and changes in endogenous polyamine content. *Plant Physiol* 70:844-848.

Galston AW (1983) Polyamines as modulators of plant development. *Bioscience* 33:382-388.

Galston AW, Kaur-Sawhney R, Tiburcio AF, Hamasaki N, Oshima T, Furuya, M (1990) The control of morphogenesis by polyamines. *In*: Polyamines and Ethylene: Biochemistry, Physiology and Interactions, Flores HE, Arteca RN, Shannon JC (eds), Am Soc Plant Physiologists, Rockville, MD, pp 224-237.

Galston AW, Flores HE (1991) Polyamines and plant morphogenesis. *In*: Biochemistry and Physiology of Polyamines in Plants, Slocum RD, Flores HE (eds), CRC Press, Boca Raton, Florida, pp 175-186.

Gamarnik A, Frydman RB (1991) Cadaverine, an essential diamine for the normal root development of germinating soybean (*Glycine max*) seeds. *Plant Physiol* 97:778-785.

Giuliano D, Rosellini D, Terzi M (1983) A new method for the purification of the different stages of carrot embryoids. *Plant Cell Rep* 2:216-218.

Grisebach H (1978) Lignins. *In*: The Biochemistry of Plants, Stumpf PK, Conn EE (eds), vol 7, Academic Press, New York, pp 457-478.

Hamill JD, Robins RJ, Parr AJ, Evans DM, Furze JM, Rhodes MJC (1990) Over-expression of a yeast ornithine decarboxylase gene in transgenic roots of *Nicotiana rustica* can lead to enhanced nicotine accumulation. *Plant Mol Biol* 15:27-38.

Hashimoto T, Yukimune Y, Yamada Y (1989) Putrescine and putrescine N-methyltransferase in the biosynthesis of tropane alkaloids in cultured roots of *Hyoscyamus albus* I Biochemical studies. *Planta* 178:123-130.

Igarashi K, Yabuk M, Yoshioka Y, Eguchi K, Hirose S (1977) Mechanism of stimulation of polyphenylalanine synthesis by spermidine. *Biochem Biophys Res Commun* 75:163-171.

Jarvis BC, Shannon PRM, Yasmin S (1983) Involvement of polyamines with adventitious root development in stem cuttings of mung bean. *Plant Cell Physiol* 24:677-683.

Jarvis BC (1986) Endogenous control of adventitious rooting in non-woody cuttings. *In*: New Root Formation in Plants and Cuttings, Jackson MB (ed), Martinus Nijhoff Publishers, Dordrecht, pp 191-222.

Kanchanapoom M, Antognoni F, Pistocchi R, Bagni N (1991) Effect of auxins on spermidine uptake into carrot protoplasts. *Physiol Plant* 82:19-23.

Kaur-Sawhney R, Tiburcio AF, Galston AW (1988) Spermidine and flower-bud differentiation in thin-layer explants of tobacco. *Planta* 173:282-284.

Litz RE, Schaffer B (1987) Polyamines in adventitious and somatic embryogenesis in mango (*Mangifera indica* L). *J Plant Physiol* 128:251-258.

Margosiak SA, Dharma A, Bruce-Carver MR, Gonzales AP, Louie D, Kuehn GD (1990) Identification of the large subunit of ribulose 1,5-bisphosphate carboxylase-oxygenase as a substrate for transglutaminase in *Medicago sativa* L (Alfalfa). *Plant Physiol* 92: 88-96.

Martin-Tanguy, J (1985) The occurrence and possible function of hydroxycinnamoyl acid amides in plants. *Plant Growth Regul* 3: 381-399.

Martin-Tanguy J, Negrel J, Paynot M, Martin C (1987) Hydroxycinnamic acid amides, hypersensitivity, flowering and sexual organogenesis in plants. *In*: Plant Molecular Biology, von Wettstein D, Chua NH (eds), Plenum Press, New York, pp 253-263.

Martin-Tanguy J, Tepfer D, Paynot M, Burtin D, Heisler L, Martin C (1990) Inverse relationship between polyamine levels and the degree of phenotypic alteration induced by the root-inducing left-hand transferred DNA from *Agrobacterium rhizogenes*. *Plant Physiol* 92:912-918.

Mengoli M, Bagni N, Luccarini G, Nuti Ronchi V, Serafini-Fracassini D (1989) *Daucus carota* cell cultures: polyamines and effect of polyamine biosynthesis inhibitors in the preembryogenic phase and different embryo stages. *J Plant Physiol* 134:389-394.

Mengoli M, Chriqui D, Bagni N (1992a) Protein, free amino acids and polyamine contents during development of hairy root *Nicotiana tabacum* plants. *J Plant Physiol* 139:697-702.

Mengoli M, Chriqui D, Bagni N (1992b) Putrescine biosynthesis and oxidation in normal and hairy root tobacco plants. *J Plant Physiol* 140:153-155.

Mengoli M, Ghelli A, Chriqui D, Bagni N (1992c) Growth kinetics, polyamine pattern and biosynthesis in hairy root lines of *Nicotiana tabacum. Physiol Plant* 85:697-703.

Mizrahi Y, Applewhite PB, Galston AW (1989) Polyamine binding to proteins in oat and *Petunia* protoplasts. *Plant Physiol* 91:738-743.

Montague MJ, Kopperbrink JW, Jaworski EG (1978) Polyamine metabolism in embryogenic cells of *Daucus carota*. I. Changes in intracellular content and rates of synthesis. *Plant Physiol* 62:430-433.

Ooms G, Bains A, Burrel MM, Karp A, Twell D, Wilcox E (1985) Genetic manipulation in cultivars of oilseed rape (*Brassica napus*) using *Agrobacterium. Theor Appl Genet* 71:325-329.

Pfosser M, Koenigshofer H, Kandeler R (1990) Free, conjugated, and bound polyamines during the cell cycle of synchronized cell suspension cultures of *Nicotiana tabacum. J Plant Physiol* 136:574-579.

Picton JM, Steer MV (1983) Evidence for the role of Ca^{2+} ions in tip extension in pollen tubes. *Protoplasma* 115:11-17.

Robins RJ, Parr AJ, Walton NJ (1991) Studies on the biosynthesis of tropane alkaloids in *Datura stramonium* L transformed root cultures. 2. On the relative contribution of L-arginine and L-ornithine to the formation of the tropane ring. *Planta* 183:196-201.

Samborski DJ, Rohringer R (1970) Abnormal metabolites of wheat: occurrence, isolation and biogenesis of 2-hydroxyputrescine amides. *Phytochemistry* 9:1939-1945.

Schiefelbein JW, Somerville C (1990) Genetic control of root hair development in *Arabidopsis thaliana. Plant Cell* 2:235-243.

Schwartz M, Altman A, Cohen Y, Arzee T (1986) Localization of ornithine decarboxylase and changes in polyamine content in root meristems of *Zea mays. Physiol Plant* 67:485-492.

Sen K, Choudhuri MM, Ghosh B (1981) Changes in polyamine contents during development and germination of rice seeds. *Phytochemistry* 20:631-633.

Serafini-Fracassini D, Bagni N, Cionini PG, Bennici A (1980) Polyamines and nucleic acids during the first cell cycle of *Helianthus tuberosus* tissue after the dormancy break. *Planta* 148:332-337.

Serafini-Fracassini D, Alessandri M (1983) Polyamines and morphogenesis in *Helianthus tuberosus* explants. *In*: Advances in Polyamine Research, Bachrach U, Kaye A, Chayen R (eds), vol 4, Raven Press, New York, pp 419-426.

Serafini-Fracassini D, Del Duca S, D'Orazi D (1988) First evidence for polyamine conjugation mediated by an enzymic activity in plants. *Plant Physiol* 87:757-761.

Serafini-Fracassini D, Del Duca S, Torrigiani P (1989) Polyamine conjugation during the cell cycle of *Helianthus tuberosus*: non enzymatic and transglutaminase-like binding activity. *Plant Physiol Biochem*, 27:659-668.

Skoog F, Schmitz R (1972) Cytokinins. *In*: Plant Physiology: a Treatise, F Steward (ed), vol 6B, Academic Press, New York, pp 181-213.

Slocum RD, Galston AW (1985) Changes in polyamine biosynthesis associated with postfertilization growth and development in tobacco ovary tissues. *Plant Physiol* 79: 336-343.

Smith MA, Davies PJ, Reid JB (1985) Role of polyamines in gibberellin-induced internode growth in peas. *Plant Physiol* 78:92-99.

Srivastava SK, Prakash V, Naik BI (1977) Regulation of diamine oxidase activity in germinating pea seeds. *Phytochemistry*16:185-187.

Stoessl A, Unwin CH (1970) The antifungal factors in barley. V. Antifungal activity of the hordatines. *Can J Bot* 48:465-470.

Tepfer D (1984) Transformation of several species of higher plants by *Agrobacterium rhizogenes*: sexual transmission of the transformed genotype and phenotype. *Cell* 37:959-967.

Tiburcio AF, Kaur-Sawhney R, Galston AW (1988) Polyamine biosynthesis during vegetative and floral bud differentiation in thin layer tobacco tissue cultures. *Plant Cell Physiol* 29:1241-1249.

Tiburcio AF, Gendy GA, Tran Thanh Van K (1989) Morphogenesis in tobacco subepidermal cells. Putrescine as marker of root differentiation, *Plant Cell Tiss Org Cult* 19:43-54.

Tiburcio AF, Kaur-Sawhney R, Galston AW (1990) Polyamine metabolism. *In*: The Biochemistry of Plants, Miflin BJ, Lea PJ (eds), vol 16, Academic Press, London, pp 283-325.

Torrigiani P, Altamura MM, Pasqua G, Monacelli B, Serafini-Fracassini D, Bagni N (1987) Free and conjugated polyamines during *de novo* floral and vegetative bud formation in thin cell layers of tobacco. *Physiol Plant* 70:453-460.

Torrigiani P, Altamura MM, Capitani F, Serafini-Fracassini D, Bagni N (1989) *De novo* root formation in thin cell layers of tobacco: changes in free and bound polyamines. *Physiol Plant* 77:294-301.

Tran Thanh Van K (1973) Direct flower neoformation from superficial tissue of small explants of *Nicotiana tabacum* L. *Planta* 115:87-92.

Tran Thanh Van K, Dien NT, Chlyah A (1974) Regulation of organogenesis in small explants of superficial tissue of *Nicotiana tabacum* L. *Planta* 119:149-159.

Vince-Prue D (1975) Photoperiodism in Plants. McGraw-Hill, New York.

Wyss-Benz M, Streit L, Ebert E (1990) Feruloylputrescine and caffeoylputrescine are not involved in growth and floral bud formation of stem explants from *Nicotiana tabacum* L var Xanthi nc. *Plant Physiol* 92:924-930.

Yang SF, Hoffman NE (1984) Ethylene biosynthesis and its regulation in higher plants. *Annu Rev Plant Physiol* 35:155-189.

crease, while exogenous application of PAs especially spermidine (Spd) and spermine (Spm) inhibits or retards the symptoms of senescence (Kaur-Sawhney *et al.*, 1982a). On the other hand, recent studies suggest a possible role of PAs in differentiation in various *in vitro* dicot systems. These include differentiation of somatic embryos in tissue cultures of carrot (Feirer *et al.*, 1984; Montague *et al.*, 1978) and *Vigna* (Kaur-Sawhney *et al.*, 1985), and differentiation of vegetative and floral buds (Flores and Protacio, 1990; Kaur-Sawhney *et al.*, 1988; Tiburcio *et al.*, 1987; Torrigiani *et al.*, 1987) as well as roots (Tiburcio *et al.*, 1989) in thin layer tobacco tissue cultures.

The information concerning the role of PAs during morphogenesis in monocot plant systems is more scarce. Nevertheless, using two *in vitro* cereal approaches, we will present in this review experimental evidence regarding the antisenescence properties of PAs as well as their relationship with the improvement of cell viability and morphogenic capacity in oat mesophyll protoplast and organogenic maize callus systems.

As *in vivo* approaches, we will examine the existence of PA gradients in different regions of roots of *Zea mays*. Using an ornamental orchid such as *Phalaenopsis*, we will also discuss correlations between PA levels and developmental patterns of different parts of flowers before and after pollination.

Regarding the mechanisms of PA action, it is difficult to evaluate which biochemical system is mainly affected during morphogenic events induced by PA treatments, since these compounds are polycations which can interact with diverse macromolecules, including DNA, RNA and proteins, and with the negative groups of membranes. In this review we will try to rationalize the available relevant data concerning the different proposed mechanisms of PA action, as well as to find criteria as to how PAs should be regarded in relation to other plant growth regulators.

EXPERIMENTAL SYSTEMS

Avena sativa Mesophyll Protoplast System

Although the regeneration of plants from protoplasts derived from embryogenic suspension cultures has been recently achieved in certain cereal systems (Rhodes *et al.*, 1988), plant regeneration from mesophyll cereal protoplasts is still one of the unsolved problems of plant cell biology. This is unfortunate because mesophyll protoplasts represent one of the best suited tools for plant genetic engineering due to their morphological and genetic uniformity, their availability in high number, and the relative absence of mutations compared with callus or suspension cultures (Vasil, 1987).

Genetic factors and injuries (excision and wounding) sustained during isolation of mesophyll protoplasts have been emphasized in the past as the main reasons for lack of sustained cell division and failure of plant regeneration in cereals (Flores *et al.*, 1981). For example, it has been suggested (Fuchs and Galston, 1976) that morphological instability and deterioration of oat mesophyll protoplasts may be related to a complex of senescence-induced changes, especially RNAse activity (Kaur-Sawhney *et al.*, 1976), following leaf excision and protoplast isolation. However, tobacco protoplasts which readily divide also show a dramatic increase in RNAse activity after 24 h of isolation (Lázár *et al.*, 1973), and this increase has been shown to be due mostly to osmotic stress rather than to injury caused during isolation of protoplasts (Premecz *et al.*, 1977).

We have formulated the hypothesis that difficulties of monocot mesophyll protoplasts in retaining viability compared with those derived from dicot protoplasts may derive from their different PA metabolism in response to the physiological stress incurred during isolation in high osmotica (Tiburcio *et al.*, 1986a,b).

With *Avena sativa* L. (oat) leaves and mesophyll protoplasts as experimental model systems, it has been shown (Tiburcio *et al.*, 1986a) that the presence of high levels of mannitol or sorbitol in the incubation media produces specific changes in the PA metabolism of dark-incubated leaves which are not produced in the absence of osmotica. Dark-incubation of oat leaves in the presence of 0.6M sorbitol results is a dramatic increase in the levels of the diamine putrescine (Put), although the levels of the triamine spermidine (Spd) and the tetraamine spermine (Spm) decrease. This response is the opposite to that observed in leaves of dicot genera, which readily regenerate plants from mesophyll protoplasts. In the solanaceous species (*Nicotiana and Capsicum*) the osmotically-induced increase in higher PAs is due mainly to Spd, whereas the legumes *Vigna* and *Trigonella* showed increased Spm levels after osmotic treatment. A similar pattern was observed when comparing freshly isolated oat protoplasts with readily regenerating dicot protoplasts (Tiburcio *et al.*, 1986a).

The high Put/Spd+Spm ratio in leaves and protoplasts of species which yield poorly dividing protoplasts contrasts with the low ratio in leaves and protoplasts of species which yield readily dividing protoplasts. This is interesting since many investigations point to an important role of the higher PAs (Spd and Spm) in DNA synthesis and mitosis (Heby, 1981). On the other hand, high concentrations of exogenously applied or endogenous Put appear toxic to plant cells (Guarino and Cohen, 1979; Strogonov *et al.*, 1972). Therefore, we suggested that this differing response of PA metabolism to osmotic stress may account in part for the failure of cereal mesophyll protoplasts to develop readily *in vitro* (Tiburcio *et al.*, 1986a).

To test this hypothesis, we attempted to prevent the accumulation of Put in oat protoplasts by pretreating oat leaves with DL-α-difluoromethylarginine (DFMA), an irreversible inhibitor of the enzyme arginine decarboxylase (ADC) responsible for the osmotically induced Put rise in cereals (Tiburcio *et al.*, 1986b). Leaf pretreatment with 10 mM DFMA before a 6 h osmotic treatment caused a small decrease of Put and a 2-fold increase of Spm levels. After 136 h of osmotic stress, Put levels in DFMA-pretreated leaves showed a small increase, but Spd levels increased dramatically (Tiburcio *et al.*, 1986b). Furthermore, the osmotically-induced inhibition of Spd-synthase activity, which catalyzes the conversion of Put to Spd, was prevented by the pretreatment of the leaves with DFMA (Tiburcio *et al.*, 1986c). The increase in the titers of Spd and Spm may be associated with the reduced chlorophyll (Chl) loss and the enhanced ability of DFMA-pretreated leaves to incorporate tritiated thymidine, uridine and leucine into macromolecules (Tiburcio *et al.*, 1986b). The higher net macromolecular synthesis and lowered rates of senescence observed in DFMA-pretreated leaves after osmotic treatment encouraged us to examine whether protoplasts isolated from such leaves will grow better in culture. As we expected, protoplasts isolated from DFMA-pretreated leaves appeared greener, with greater cytoplasmic streaming and more uniform regeneration of cell walls, and showing less lysis than protoplasts extracted from control leaves. The overall viability and cell division of the protoplasts was significantly improved by the DFMA-pretreatment, thus supporting the hypothesis that the osmotically induced rise in Put and blockage of its conversion to Spd and Spm may contribute to the lack of sustained cell division in cereal mesophyll protoplasts (Tiburcio *et al.*, 1986b).

The DFMA-pretreatment, which increases the endogenous levels of Spd and Spm, was clearly superior in improving oat mesophyll protoplast culture than previously reported treatments using exogenous applications of PAs (Kaur-Sawhney *et al.*, 1980).

It has been recently reported that oat mesophyll protoplasts are capable of proliferation in *in vitro* culture and even able of microcolony formation when protoplasts are grown in the presence of *Triticum monococcum* feeder cultures (Hahne *et al.*, 1989). These results are similar to those first reported by us (Tiburcio *et al.*, 1986b).

In our hands, the mesophyll oat protoplasts pretreated with DFMA were also stable, not undergoing the phenomenon of lysis often encountered with cereal mesophyll protoplasts, and many of them were able to regenerate complete cell walls (Fig. 1), as opposed to the usual patchy deposition of cellulose (Kinnersley *et al.*, 1978). One-month old DFMA-pretreated cultures showed an increase in all stages of protoplast development when compared with control protoplasts (Tiburcio *et al.*, 1986b). While control cultures had all senesced, most of the DFMA-pretreated cultures were maintained in liquid droplets for up to 2 months (Tiburcio *et al.*, 1986b). Such cultures when transferred to solid media were also capable of microcolony formation (Fig. 1).

In current experiments we are assaying different treatments and conditions in order to further improve the cell division of oat mesophyll protoplasts. One of the approaches has been the use of guazatine, an inhibitor of PA oxidase (PAO) activity, with the aim to prevent the degradation of endogenous Spd and Spm levels in osmotically-stressed oat leaves incubated in the dark. The results have shown that guazatine treatments are able to decrease the levels of diaminopropane (Dap), with a concomitant increase of the levels of Spd and Spm. This effect on PA metabolism is positively correlated to the capability of guazatine to prevent Chl loss and integrity of the leaves after dark-incubation on high osmotica (Capell *et al.*, 1993). We are now interested in knowing whether or not guazatine treatments are able as well to improve the viability and cell divison of oat mesophyll protoplasts.

Zea mays Organogenic-Callus Culture System

Until about 1980 there were only a few reports of reproducible plant regeneration from cereal tissue cultures. In most instances even the induction and maintenance of long-term callus cultures were found extremely difficult. In those species where plant regeneration was possible, it was limited to a few genotypes, was sporadic and transient in nature, and the number of plants formed was small. The recent use of 2,4-dichlorophenoxy acetic acid (2,4-D) and culture of explants obtained from tissues or organs containing cells that are not irreversibly committed to a specific developmental pattern, has allowed the improvement of plant regeneration from cereal tissue cultures, either via organogenesis or via somatic embryogenesis (Vasil, 1987).

However, variations still exist among genotypes and even within tissues produced by the same genotype. Other factors which may contribute to these variations are the de-

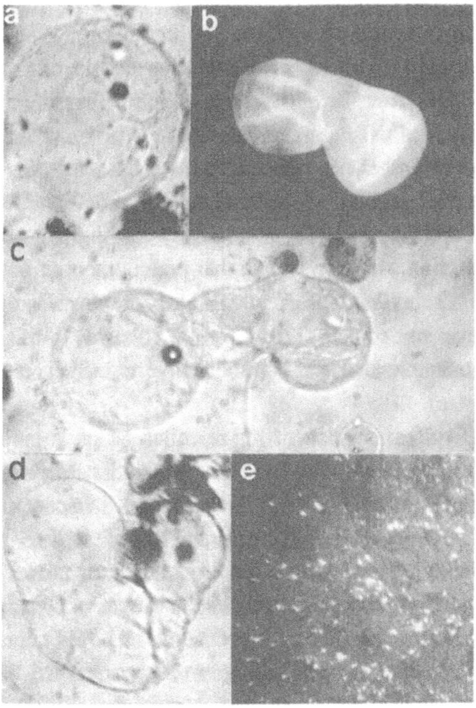

Figure 1. Development of oat mesophyll protoplasts pretreated with DFMA. (a) 'Oval' stage. (b) 'Double' cell stained with calcofluor white: note that freshly formed cell wall is clearly visible. (c) 'Triple'. (d) 'Multiple', clusters of cells. (e) Microcolonies formed when DFMA-pretreated cultures maintained in liquid droplets for up to 2 months were transfered to solid media (Tiburcio, Kaur-Sawhney, and Galston, 1986).

velopmental stage of the original explant, the age of the cultures, medium composition and environmental conditions (Duncan *et al.*, 1985; Vasil, 1987). For instance, it has been observed that even those genotypes which are considered highly recalcitrant can be induced to undergo morphogenesis when suitable explants at defined developmental stages are excised from plants grown under optimal conditions. Conversely, genotypes that normally undergo morphogenesis in culture are difficult or impossible to regenerate from explants that are otherwise unsuitable and are obtained from plants growing under adverse conditions (Vasil, 1987). Morphogenesis must therefore be considered a complex phenomenon regulated by the interaction of genotype with physiological and environmental factors, which in turn influence the synthesis, transport and the availability of plant growth regulators.

The endogenous pools of plant growth regulators have been causally related to morphogenic competence of tissues cultured *in vitro* (Vasil, 1987). In this regard, PAs can be included among the internal factors controling plant morphogenesis, since many correlations have been established, mainly with dicot cells and tissue culture systems. Thus, changes in endogenous levels of PAs have been correlated with the induction of somatic embryos in

Orchid flowers are highly modified and differ from other flowers in several details. In monandrous orchids (which constitute most of the family) the stamens, stigmas and styles are fused into one organ, the gynostemium or column, which bears a simple fertile stamen near or at its apex. In many species the sepals and petals often appear to be almost identical. The dorsal sepal may be slightly different in some species and the median petal is often modified considerably to form a showy (often gaudy) labellum or lip. In the bud stage the column is generally positioned below the labellum, but as the flower opens it undergoes resupination (i.e. twisting around) and the relative positions of these floral parts are reversed (Arditti, 1984).

Orchid flowers, if unpollinated, remain alive for long periods of time. Some flowers, which stay fresh for up to 3-4 months in the absence of pollination, start to show signs of senescence or exhibit typical post-pollination phenomena quickly (in some orchids within 15 min) after being pollinated (Arditti, 1979). Depending on species or genus, pollination can cause a variety of events including wilting, senescence, greening, fading, or anthocyanin formation in some or all parts of the flower (Arditti, 1979).

Serious attempts have been made not only to manipulate and control flower senescence in orchids, but also to gain more information concerning the physiology of this process. In this regard, much attention has been given to the role of plant hormones. For instance, it is known that auxins are able to mimic certain pollen effects and initiate a number of post-pollination phenomena in many orchids (Arditti, 1979). Ethylene is also a negative effector which has been implicated in the control of several post-pollination processes. Thus, wilting and senescence of various parts of orchid flowers produced after pollination have been associated with the presence of ethylene (Chadwick *et al.*, 1980; Hew *et al.*, 1989). On the contrary, it seems that cytokinins may be involved in delaying flower senescence. In fact, endogenous cytokinins have been suggested as being the natural antisenescence factors in orchid flowers (Arditti *et al.*, 1971).

In the previous sections, we have already shown the antisenescence properties of PAs on cell and tissue culture development in monocots. Here we report the existence of specific PA patterns during post-pollination development of different parts of *Phalaenopsis* flowers.

In unpollinated flower, the PA pattern of different flower parts are quite specific (see Fig. 2). Petals and sepals as well as the column contain Spd as the main PA, showing also significant amounts of Spm. In contrast, the levels of Spd and Spm are very low or almost negligible in ovary and labellum, while the levels of Dap (an oxidation product of Spd and Spm) reach the highest values (Fig. 2).

Once the flower is pollinated, the levels of Spd in petals and sepals show a transient increase 1 day post-pollination, but then follow a dramatic decrease from the 3rd to the 7th day. In contrast, the levels of Dap show the highest values after 1 week post-pollination (Fig. 2). In the column, the levels of Spd and Spm dramatically decrease 1 day post-pollination. The total PA levels reach the highest value on the third day, but then decrease at day 7 after pollination (Fig. 2). In the labellum, total PA levels show a transient increase 1 day post-pollination, but then they decrease on the 3rd and 7 day after pollination (Fig. 2). The ovary is the only part of the orchid flower which shows a linear increase of all PAs upon pollination (Fig. 2).

Although further work is needed, these data are consistent with the view that PAs (especially Spd and Spm) may be included among the antisenescence factors responsible for

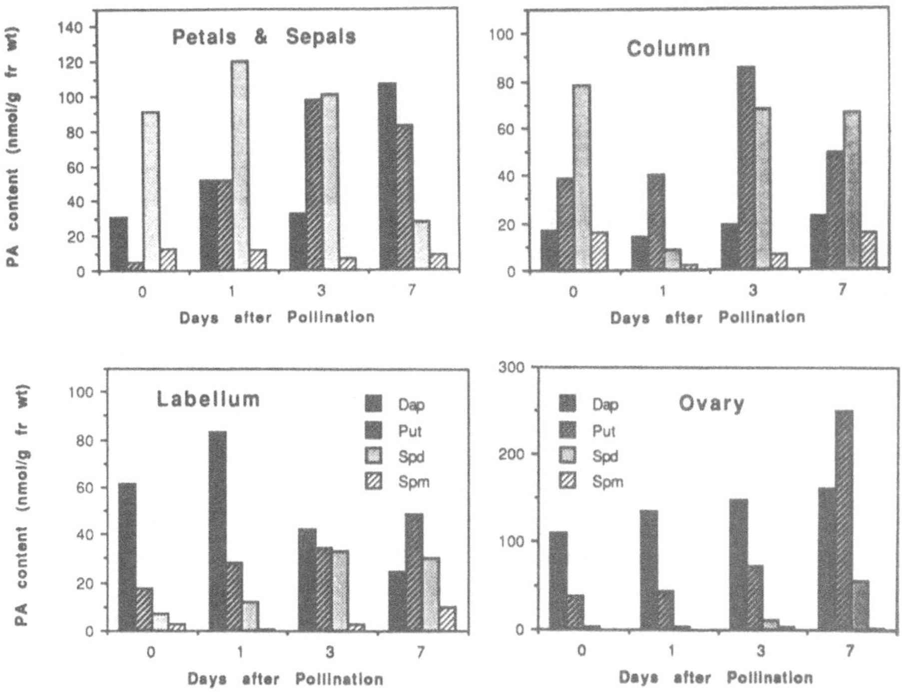

Figure 2. Polyamine content in *Phalaenopsis* flower parts during a 7-day post-pollination period.

the maintenance of the quality of petals and sepals of unpollinated *Phalaenopsis* flowers. Once the flower is pollinated, PA changes are well correlated with morphological and developmental modifications of each flower part.

MECHANISMS OF ACTION

Much of the work presented in the previous sections has involved the development of firm correlations between some aspects of PA metabolism and the morphogenic behaviour in monocot systems of mesophyll protoplasts, organogenic callus and specific plant organs. Thus, the fact that Spd and/or Spm contents decline during aging of *in vitro* cultured protoplasts and tissues or during senescence of intact leaves and flowers, together with the antisenescent action of applied Spd or Spm, leads to an association of high PA levels with continuing cellular integrity and delay of aging-senescence. The existence of gradients of Spd and Spm correlated with meristematic activity along the main root and lateral root primordia of maize gives further support to the above contention.

We have also shown that pretreatment of oat leaves with DFMA followed by osmotic incubation, routinely used for protoplast isolation, leads to an enhancement of endogenous Spd and Spm levels, which in turn results in antisenescence effects similar to those induced by exogenous supply of these PAs. During rejuvenation of previously DFMA-treated maize callus, there is also an increase of endogenous PA levels, and this is correlated with improved plant regeneration.

All these results reinforce the view that exogenous PAs, as their endogenous counterparts may have specific functions during aging-senescence processes accompanying loss of morphogenic competence in *in vitro* cultured or intact monocot plant systems.

The phrase "the function of the PAs is unknown" or some variation thereof, is frequently encountered, but this is an overcautious statement. The fact is that many functions of PAs are known and significant progress has been made, at least in animal systems, toward an understanding of the molecular mechanisms that relate these actions to cellular functions. Since PAs are polycations which can interact with diverse macromolecules of polyanionic nature, much progress has been made in a detailed understanding of such interactions and of the molecular mechanisms involved (see review in Marton and Morris, 1987). In plants, although much less is known, it has been suggested that most of these functions and interactions of PAs with macromolecules and cellular structures may also exist (reviewed by Slocum *et al.*, 1984).

Because of such a variety of possible biological functions, it is difficult to evaluate which biochemical system is mainly affected during specific plant morphogenic events induced by PA treatments. One way to rationalize this problem could be to keep in mind the different approaches used to investigate hormone action. In this case, it has been hypothesized that the interaction of hormones with specific membrane receptors produces a primary response, which, through a complex chain of events, leads to a physiological response (Napier and Venis, 1990). Accordingly, the search for membrane receptors and the study of primary responses involving changes in gene expression, thus affecting macromolecular synthesis,. should be neccessary steps in understanding the mechanisms of hormone action. In addition, the understanding of plant growth hormonal effects on the levels and metabolism of well recognized second messengers will help to delineate in the future the definitive signal transduction pathways.

Membrane Level

The finding that small amounts of Spm could stabilize osmotically fragile bacteria was the earliest evidence that PAs may interact with either the cell wall or the cell membrane (see reviews in Bachrach, 1973; Tabor and Tabor, 1964). The same effect was found for the protection of bacterial protoplasts by Tabor (1960), who showed that interaction of PAs with the cell membrane was involved in its stabilization. Grossowicz and Ariel (1963) indicated that Spd did not interfere with the action of lysozyme, used in the protoplast isolation, but it did prevent the leakage of intracellular constituents. Although other organic cations produced the same effect as Spm, Mg^{+2} did not protect protoplasts from lysis.

Tabor and Tabor (1964) pointed out that, compared with the concentration of sucrose usually employed (0.5 M), the low concentration of Spm (1 mM) necessary to prevent protoplast lysis indicates that Spm does not act through an osmotic mechanism as has been proposed for sucrose. They postulated that Spm forms complexes with acidic groups in the cell membrane, which reduces repulsive forces without diminishing cohesive forces and thereby stabilizes the membrane.

More recent studies have attempted to characterize the interactions of PAs with cellular membranes in mammalian systems (see review in Marton and Morris, 1987). For instance, Ballas *et al.* (1983) showed that PAs decreased erythrocyte deformability and stabilized the membranes through an association with cytoskeletal proteins on the cytoplasmic surface of the membranes. In addition, they showed that Ca^{+2} or Mg^{+2} could not replace the PAs, thus suggesting that charge *per se* was not responsible for the observed effects. Other studies have postulated that competition between PA biosynthetic inhibitors and PAs for low-affinity binding sites may occur at the outer surface of the inner mitochondrial membrane. A similar type of competition for membrane receptors between DFMO and testosterone has also been reported (review in Marton and Morris, 1987)

In plants, physiological concentrations of Spd and Spm have been reported to stabilize oat mesophyll protoplasts (Altman *et al.*, 1977) and thylakoid membranes of barley chloroplasts (Popovic *et al.*, 1979) as well as to prevent Chl loss in senescing leaves of barley, radish and *Hydrilla verticillata* (Chen and Kao, 1986; Kar and Choudhuri, 1986; Srivastava *et al.*, 1983). In a previous section, we have already described the ability of DFMA-pretreatment in stabilizing oat mesophyll protoplasts against lysis and improving cell division (Tiburcio *et al.*, 1986b). The ability of guazatine in preventing Chl loss in senescing leaves of oat (Capell *et al.*, 1993) has also been described.

Polyamine binding is also known to affect membrane fluidity and thus PAs may modulate the activities of membrane-associated enzymes indirectly (review in Slocum *et al.*, 1984). For example, the ethylene-synthesizing complex in plants, which was thought to be associated with the plasma membrane, was inhibited by Spm in apple tissue. The inhibition of ethylene biosynthesis was related to the prevention by Spm of decreased membrane microviscosity associated with senescence and, possibly, with resultant conformational changes in the ethylene-synthesizing complex (Apelbaum *et al.*, 1981).

Using microsomal membranes from *Phaseolus*, Roberts *et al.* (1986) reported that exogenously applied PAs are able to interact with membrane lipids, thereby substantially reducing membrane fluidity. The authors caution that some physiological effects attributed to exogenous application of PAs may thus reflect artificially induced membrane rigidification. They also indicated that the antisenescence effects of PAs are not unlike those of Ca^{+2}, which is thought to delay the onset of senescence in particular plant systems by rigidifying membranes (Roberts *et al.*, 1986).

However, the interpretations of this last study should be undertaken with some caution as well. For instance, although the effects of 1 mM Spm and Spd (concentrations normally reported to elicit physiological effects) on polarization values were statistically significant, the increases on such values were only 10% and 8%, respectively. On the other hand, it is not possible to make a general association between the antisenescence effects of PAs with those reported for Ca^{+2}, since antagonistic effects have been attributed to these compounds in various senescing plant systems. Thus, Kaur-Sawhney and Galston (1979) have shown that Ca^{+2} interferes with the antisenescence effects of Spm during dark-induced senescence of oat leaves. Further evidence based on studies using peeled oat leaves (Fuhrer *et al.*, 1982) and apple tissue (Apelbaum *et al.*, 1981) suggest the existence of a competition between PAs and Ca^{+2} for the same membrane binding site (reviewed by Kaur-Sawhney and Galston, 1991).

In a recent investigation, we have analyzed the effects of exogenous PA supply and

lational machinery where PAs may be acting. Thus it is well known that PAs bind to ribosomes facilitating the association and stability of ribosomal subunits.There exists also evidence about the influence of PAs on translational fidelity (the frequency with which aminoacids are correctly incorporated into a polypeptide in response to a given codon) and on chain termination. Polyamines may also affect either the initiation step or the elongation of nascent polypeptide chains. A number of aminoacyl tRNA synthetases have been reported to be stimulated by PAs. In addition to influencing binding of the tRNA to the homologous synthetase, structural transitions induced by PAs can influence codon-anticodon interaction with the ribosome (reviews in Marton and Morris, 1987; Tabor and Tabor, 1984).

A variety of studies have reported the influence of PAs on the synthesis of specific proteins (review in Marton and Morris, 1987). In general, PAs have been found to stimulate the synthesis of high-molecular-weight proteins to a greater extent rather than stimulation of the synthesis of low-molecular-weight proteins.

Post-translational modifications of proteins involving PAs are also known. Perhaps the most studied is the mechanism by which PAs may become bound to proteins by the action of transglutaminases (review in Pegg, 1986). Other post-translational mechanisms (not mediated by transgutaminases) such as PA binding to proteases and PA-mediated phosphorylation of nuclear proteins are also known.

Most of the work regarding the antisenescence effects of PAs on macromolecular synthesis in plants has been performed using oat mesophyll protoplast and oat leaf systems (described in previous section). Several patterns related to macromolecular metabolism quickly emerged from protoplasts experiments. While incorporation of H^3-leucine into protein and H^3-uridine into RNA proceeds vigorously during the initial 6 to 8 h after protoplast isolation, the rate of both processes then decreases rapidly. Shortly thereafter, protein, RNA and Chl levels decline and the cells go into a senescent state that culminates in lysis. The synthesis of DNA, as evaluated by H^3-thymidine incorporation, appears to be blocked in the protoplast, which shows only sporadic mitotic activity. Application of exogenous PAs stabilize oat protoplasts against lysis, increase rates of protein and RNA synthesis, and initiate DNA synthesis and mitotic activity (Altman *et al.*, 1977; Galston *et al.*, 1978).

Osmotically-stressed oat leaves incubated in the dark show similar changes to that observed in protoplasts extracted from such leaves. Incorporation of H^3-labelled thymidine and leucine into macromolecules was reduced by more than 50%, while H^3-uridine incorporation was reduced by more than 25% after 6 h of osmotic stress. Leaf pretreament with DFMA followed by incubation on 0.6M sorbitol increased the incorporation of all the precursors into macromolecules, possibly as a consequence of increased endogenous PA levels (Tiburcio *et al.*, 1986b).

In dark-incubated oat leaves and protoplasts, a correlation exists between the decline in RNA and protein levels and the rise in the activities of several hydrolytic enzymes, especially RNase and protease (Galston *et al.*, 1978; Kaur-Sawhney *et al.*, 1978). The rise of both hydrolytic enzyme activities is prevented by exogenous application of PAs (Kaur-Sawhney *et al.*, 1978).

In the case of RNase, recent experiments have suggested that both *de novo* synthesis and enzyme activity are inhibited by PAs, although non-specific and/or indirect effects should not be discarded (Isola and Franzoni, 1989).

Much more is known about the mode of PA action on protease activity (Kaur-Sawhney *et al.*, 1982b). The dramatic rise in protease activity caused by excision and dark incubation of oat leaves involves both an acid and a neutral form of protease. The acid protease increased more than 6-fold, while the neutral protease increased about 2.5-fold during a 48h dark incubation. This rise in both protease activities was inhibited by floating leaves on PA solutions. The most effective PA was Spm, followed by Spd, Put and cadaverine.

To understand the mechanism of PA action on protease activity, Spd was preincubated in a cell-free system with the enzyme or substrate preparations. Pretreatment of the enzyme with Spd prior to addition of substrate inhibited both the protease activities, while pretreatment of the substrate with Spd did not (Kaur-Sawhney *et al.*, 1982b). These data support the view that Spd inhibits protease activity by interacting with the enzyme and not the substrate. This interaction of Spd with the enzyme probably does not result from the transglutaminase-mediated formation of covalent linkages, since it is insensitive to EDTA. In addition, it appears that such interaction may involve ionic binding, since Spd effect is prevented by Ca^{+2} (Kaur-Sawhney *et al.*, 1982b).

A direct interaction between the negative sites on the enzyme and the positively charged PAs by means of electrovalent linkages has been recently proposed using purified alfalfa leaf protease in a cell-free system (Balestreri *et al.*, 1987). These authors have shown that Spm interacts with the enzyme molecule at a site different from the substrate binding site, which results in a transition of the enzyme from a catalytically fully active conformational state to a half-active one. The electrovalent binding and the consequent transition of the enzyme to the half-active state occurs only at low ionic strength, but is prevented by high ionic strength. Accordingly, Balestreri *et al.* (1987) have proposed a novel control mechanism of plant proteolytic enzymes. In senescing leaves the shortage of water, and the resulting increase in ionic strength, prevents the binding of the naturally occurring polycations, including PAs, to the proteolytic enzymes thus maintaining the later in the fully active conformation state. Conversely, during normal plant growth the ionic strength is low thus facilitating the PA binding and the subsequent inactivation of proteolytic enzymes (Balestreri *et al.*, 1987). Unfortunately, the physiological significance of such mechanisms has not been assessed *in vivo*.

Senescence usually occurs in plants when ethylene, a gaseous plant hormone, is released in relatively large quantities. Ethylene arises from SAM, through ACC as an intermediate. Thus, SAM sits astride a critical branch point in plant metabolism. Commitment toward ACC causes a senescence syndrome characterized by increased RNase, protease and Chl loss, while a commitment toward aminopropyl transfer and Spd synthesis not only prevents senescence, but continues the active anabolic and growth activities of the plant. These pathways are competitive, for when specific inhibitors are used to prevent either ethylene or Spd biosynthesis, the flow of carbon atoms from SAM through the alternative pathway is favoured. Furthermore, each end product inhibits the formation of the other; thus, once Spd synthesis is favoured, ethylene evolution is progressively diminished, and *vice versa*. These observations raise the possibility that plant senescence is controlled, in some measure, by interaction between PAs and ethylene (see recent review in Kushad and Dumbroff, 1991).

In the previous section, we have indicated that the mechanisms by which PAs modulate senescence and the biosynthesis of ethylene presumably include the stabilization of membranes. In addition to these effects at the membrane level, we will discuss in this section recent evidence indicating that exogenous PAs are also able to control ethylene biosynthesis at the level of ACC synthase transcript accumulation (Li *et al.*, 1992).

Previous publications showed that PAs were able to inhibit the development of ACC synthase activity and/or ethylene production in various plant senescing systems (see review in Kushad and Dumbroff, 1991). This raises the question of whether the PA effect involves the regulation of the ACC synthase transcript level. To address this question, regulation of wound-inducible ACC synthase expression as affected by PA treatment in tomato fruits has been studied (Li *et al.*, 1992). First, cell-free extracts were prepared to determine whether or not the production of ACC synthase activity upon wounding was inhibited by PAs. The results were consistent with previous reports showing that Spm has a stronger inhibitory effect on both ACC synthase activity and ethylene production than other PAs. The levels of the wound-inducible ACC synthase transcript in samples incubated for 1 to 6 h following wounding in the presence or absence of PAs were analyzed on northern blots. The results showed that the different PAs inhibited the accumulation of the wound-inducible ACC synthase transcript to different degrees. As in the case of the ACC synthase activity, the inhibition by Spm was greater than that caused by Spd or Put alone. The specificity of the PA effect on the ACC synthase transcript level was tested by hybridizing the blots separately to probes for a wound-repressible glycine-rich protein gene and a constitutively expressed rRNA. The results showed that the patterns of these transcripts were not affected as markedly by PAs. Taken together, these data implicate PAs as regulators of wound-induced ACC synthase transcript (Li *et al.*, 1992).

However, at this time it is not known whether the increase in ACC synthase transcript level results from an increase in transcription rate or from an increase in mRNA stability, or from both. Nuclear run-on transcription assays need to be performed to differentiate between the various possible mechanisms involved in the regulation of ACC synthase gene expression by PAs.

Using the *Avena sativa* leaf-system, we have performed experiments attempting to address the question of whether or not changes in ADC activity induced by osmotic treatment are accompanied by parallel changes of ADC transcript level.

Previous work on the same system showed that excised oat leaf pieces responded to osmotic stress within a few minutes by a marked rise in ADC activity and Put content (Flores and Galston, 1982). Through the use of cycloheximide, it was shown that protein synthesis was necessary for this response to occur, and through the use of DFMA it was possible to establish that ADC activity is responsible for the rise in Put level. The enzyme ADC was subsequently purified from stressed oat leaves and it migrated as a single band on denaturing gels, with molecular weight of about 39 KD. By measuring ^{35}S-methionine incorporation into proteins of stressed and unstressed leaves, it was shown that stress increased the amount of the putative ADC at the MW 39,000 location by about 30% (Flores *et al.*, 1985). From all these results, the authors suggested that *de novo* synthesis of ADC could be involved in the plant response to osmotic stress, although more precise details concerning the level of regulation of ADC synthesis could not be investigated.

The availability of a cDNA clone encoding oat ADC (Bell and Malmberg, 1990), has now permitted us to study this problem in greater detail. The levels of ADC transcript in oat leaf-samples incubated on 0.6 M sorbitol were analyzed by northern blots and compared with those of unstressed leaf tissue. Preliminary results show that there is no significant increase of ADC transcript if the incubations on high osmotica were either done in the dark or in the light (Campos *et al.*, submitted). Also a similar situation has been found in animal systems in which increased synthesis of ODC protein was not accompanied by any change in the levels of ODC transcript (Persson *et al.*, 1985). It is possible that the regulation by osmotic stress of the ADC gene expression could be exerted at the level of ADC mRNA translation, although further experiments should be performed to confirm this hypothesis.

In connection with the antisenescence properties of exogenous PAs in osmotically-stressed oat leaves, we have initiated experiments to investigate if PAs, in addition to acting at the membrane level, are also able to influence ADC gene expression. Preliminary results show that there is an increase of the ADC transcript in oat leaves incubated on high osmotica in the presence of Spm. However, this effect is not accompanied by parallel changes in ADC activity (Campos *et al.*, submitted). Further experiments are in progress in order to understand the mechanism involved in the regulation of ADC gene expression by PAs.

It is already known that PAs can induce post-translational modifications of specific proteins through covalent linkage (probably mediated by transglutaminase) in several plants, including monocot systems, such as cultured oat mesophyll protoplasts and rice cell suspensions (Mehta *et al.*, 1991; Mizrahi *et al.*, 1989). Therefore, it is conceivable that this mechanism may also have some physiological function regarding the antisenescence properties of PAs (Mizrahi *et al.*, 1989).

Signal Transduction Mechanisms: Hormonal versus PA Effects

The molecular mechanisms by which plants convey signals in response to environmental and hormonal stimuli are largely *terra incognita*. Our aim in this section is only to compare hormonal effects with those produced by PAs on the turnover of inositolphospholipids, which involves a well known signal transduction mechanism (Berridge, 1987). Thus, extensive work in animals has shown that the hydrolysis of inositolphospholipids produce two second messengers, inositol 1,4,5-triphosphate (IP_3) and diacylglycerol, and now there is growing evidence that this signalling mechanism involving the phosphatidylinositol (PI) cycle may function in plants as well. Polyphosphoinositides, phosphatidylinositol 4-phosphate (PIP) and phosphatidylinositol 4,5-bisphosphate (PIP_2) have been shown to be present in a number of plants. It has also been shown that isolated plant membranes including maize microsomes and oat membrane vesicles exhibit appropriate responses (i.e. increased Ca^{+2} efflux) to exogenous applied IP_3 (see references in Dureja-Munjal *et al.*, 1992).

Plant hormones might utilize the PI cycle for the transduction of signals between plant cells. This contention is supported by effects on PI metabolites and/or PI turnover in response to auxins, cytokinins and gibberellic acid (see references in Dureja-Munjal *et al.*, 1992).

With regard to PAs, interaction between Spm and membrane phospholipids has

Capell T, Campos JL, Tiburcio AF (1993) Antisenescence properties of guazatine in osmotically-stressed oat leaves. *Phytochemistry*, in press.

Chadwick A, Hogan N, Arditti J (1980) Post-pollination phenomena in orchid flowers. IX. Induction and inhibition of ethylene evolution, anthocyanin synthesis, and perianth senescence. *Bot Gaz* 141: 422-427.

Chen C, Kao C (1986) Localized effect of 1,3-diaminopropane and benzyladenine on chlorophyll loss in soybean primary leaves. *Bot Bull Acad Sin* 27: 97-100.

Cocucci S, Bagni N (1968) Polyamine induced activation of protein synthesis in ribosomal preparation from *Helianthus tuberosus* tissue. *Life Sci* 7: 113-120.

Drolet G, Dumbroff EB, Legge RL, Thompson JE (1986) Radical scavenging properties of polyamines. *Phytochemistry* 25: 367-371.

Dumbroff E (1991) Mechanisms of polyamine action during plant development *In*: Lecture Course on Polyamines as Modulators of Plant Development, Galston AW, Tiburcio AF (eds), vol 257 Fundación Juan March, Madrid, pp 62-66.

Dumortier FM, Flores HE, Shekhawat NS, Galston AW (1983) Gradients of polyamines and their biosynthetic enzymes in coleoptiles and roots of corn. *Plant Physiol* 72: 915-918.

Duncan D, Williams M, Zehr B, Wildholm J (1985) The production of callus capable of plant regeneration from immature embryos of numerous *Zea mays* genotypes. *Planta* 165: 322-332.

Dureja-Munjal I, Acharya M, Guha-Mukherjee S (1992) Effect of hormones and spermidine on the turnover of inositolphospholipids in *Brassica* seedlings. *Phytochemistry* 31: 1161-1163.

Feirer RP, Mignon, Litvay JD (1984) Arginine decarboxylase and polyamines required for embryogenesis in the wild carrot. *Science* 223: 1433-1435.

Flores HE, Galston AW (1982). Polyamines and plant stress: Activation of putrescine biosynthesis by osmotic shock. *Science* 211: 1259-1261.

Flores HE (1983) Studies on the physiology and biochemistry of polyamines in higher plants PhD Dissertation Yale University.

Flores HE, Kaur-Sawhney R, Galston AW (1981) Protoplasts as vehicles for plant propagation and improvement *In* : Advances in Cell Culture, Maramorosch K (ed), vol 1, Academic Press, New York, pp 241-279.

Flores HE, Protacio C (1990) Polyamine metabolism in plant cell and organ culture *In:* Polyamines and Ethylene: Biochemistry, Physiology, and Interactions, Flores HE, Arteca RN, Shannon JC (eds), American Society of Plant Physiologists, Rockville, Maryland, pp 126-137.

Flores HE, Young, ND, Galston, AW (1985) Polyamine metabolism and plant stress *In:* Cellular and Molecular Biology of Plant Stress, Key JL, Kosuge T (eds), Alan R Liss, New York, pp 93-114.

Flores HE, Protacio C, Signs M (1988) Primary and secondary metabolism of polyamines in plants *In*: Plant Nitrogen Metabolism, Poulton JE, Romeo JT, Conn EE (eds), vol 23, Plenum Press, New York, pp 329-393.

Fuchs Y, Galston AW (1976) Macromolecular synthesis in oat leaf protoplasts. *Plant Cell Physiol* 17: 475-482.

Fuhrer J, Kaur-Sawhney R, Shih L-M, Galston AW (1982) Effect of exogenous 1,3-diaminopropane and spermidine on senescence of oat leaves II Inhibition of ethylene and possible mode of action. *Plant Physiol* 70: 1597-1600.

Galston AW, Altman A, Kaur-Sawhney R (1978) Polyamines, ribonuclease and the improvement of oat leaf protoplasts. *Plant Sci Lett* 11: 69-79.

Galston AW, Kaur-Sawhney R (1982) Polyamines: are they a new class of growth regulators? *In:* Plant Growth Substances, Wareing P (ed), Academic Press, London, pp 451-461.

Gelein C (1984) Catalogue: cut flowers-pot plants-bedding plants *In:* Verenige Bloemenveilingen Aalsmeer, The Netherlands, pp 105-115.

Grossowicz N, Ariel M (1963) Mechanism of protection of cells by spermine against lysozyme-induced lysis. *J Bacteriol* 85: 293-300.

Guarino L, Cohen S (1979) Mechanism of toxicity of putrescine in Anacystis nidulans. *Proc Natl Acad Sci USA* 76: 3660-3664.

Hahne B, Fleck J, Hahne G (1989) Colony formation from mesophyll protoplasts of a cereal oat. *Proc Natl Acad Sci USA* 86: 6157-6160.

Heby O (1981) Role of polyamines in the control of cell proliferation and differentiation. *Differentiation* 19: 1-20.

Hew C, Tan S, Chin T, Ong T (1989) Influence of ethylene on enzyme activities and mobilization of materials in pollinated *Arachnis* orchid flowers. *J Plant Growth Regul* 8: 121-130.

Isola MC, Franzoni L (1989) Inhibition of net synthesis of ribonuclease by polyamines in potato tuber slices. *Plant Sci* 63: 39-45.

Kar RK, Choudhuri MA (1986) Effects of light and spermine on senescence of *Hydrilla* and spinach leaves. *Plant Physiol* 80: 1030-1033.

Kaur-Sawhney R, Rancillac M, Staskawicz B, Adams W, Galston AW (1976) Effect of cycloheximide and kinetin on yield, integrity and metabolic activity of oat leaf protoplasts. *Plant Sci Lett* 7: 57-67.

Kaur-Sawhney R, Altman A, Galston AW (1978) Dual mechanisms in polyamine-mediated control of ribonuclease activity in oat leaf protoplasts. *Plant Physiol* 62: 158-160.

Kaur-Sawhney R, Galston AW (1979) Interaction of polyamines and light on biochemical processes involved in leaf senescence. *Plant Cell Environ* 2: 189-196.

Kaur-Sawhney R, Flores HE, Galston AW (1980) Polyamine-induced DNA synthesis and mitosis in oat leaf protoplasts. *Plant Physiol* 65: 368-371.

Kaur-Sawhney R, Shih L, Flores H, Galston AW (1982a) Relation of polyamine synthesis an titer to ageing and senescence in oat leaves. *Plant Physiol* 69: 405-410.

Kaur-Sawhney R, Shih L, Cegielska T, Galston AW (1982b) Inhibition of protease activity by polyamines Relevance for control of leaf senescence. *FEBS Lett* 145: 345-349.

Kaur-Sawhney R, Shekhawat NS, Galston AW (1985) Polyamine levels as related to growth, differentiation and senescence in protoplast-derived cultures of *Vigna aconitifolia* and *Avena sativa*. *Plant Growth Regul* 3: 329-337.

Kaur-Sawhney R, Tiburcio AF, Galston AW (1988) Spermidine and flower-bud differentiation in thin layer explants of tobacco. *Planta* 173: 282-284.

Kaur-Sawhney R, Galston AW (1991) Physiological and biochemical studies on the anti-senescence properties of polyamines in plants *In:* Biochemistry and Physiology of Polyamines in Plants, Slocum RD, Flores HE (eds), CRC Press, Boca Raton, pp 201-211.

Kinnersley A, Racusen R, Galston AW (1978) A comparison of regenerated cell walls in tobacco and cereal protoplasts. *Planta* 139: 155-158.

Kushad M, Dumbroff E (1991) Metabolic and physiological relationships between the polyamine and ethylene biosynthetic pathways *In:* Biochemistry and Physiology of Polyamines in Plants, Slocum RD, Flores HE (eds), CRC Press, Boca Raton, pp 77-92.

Lázár G, Borbély G, Udvardy J, Premecz G, Farkas G (1973) Osmotic shock triggers an increase in ribonuclease level in protoplasts isolated from tobacco leaves. *Plant Sci Lett* 1: 53-57.

Li N, Parsons B, Liu D, Mattoo A (1992) Accumulation of wound-inducible ACC synthase transcript in tomato fruit is inhibited by salicylic acid and polyamines. *Plant Mol Biol* 18: 477-487.

Marton L, Morris D (1987) Molecular and cellular functions of the polyamines *In:* Inhibition of Polyamine Metabolism, McCann PP, Pegg A, Sjoerdsma A (eds), Academic Press, San Diego, pp 79-105.

Mehta A, Saftner R, Schaeffer G, Matto A (1991) Translational Modification of an 18 kilodalton polypeptide by spermidine in rice cell suspension cultures. *Plant Physiol* 95: 1294-1297.

Mizrahi Y, Applewhite PB, Galston AW (1989) Polyamine binding to proteins in oat and *Petunia* protoplasts. *Plant Physiol* 91: 738-743.

Montague MJ, Koppenbrink JW, Jaworski EG (1978) Polyamine metabolism in embryogenic cells of *Daucus carota*. *Plant Physiol* 62: 430-433.

Morris D, Jordstad C (1973) Growth and macromolecular composition of a mutant of *Escherichia coli* during polyamine limitation. *J Bacteriol* 113: 271-277.

Napier R, Venis M (1990) Receptors for plant growth regulators: recent advances. *J Plant Growth Regul* 9: 113-126.

Pegg A (1986) Recent advances in the biochemistry of polyamines in eukaryotes. *Biochem J* 234: 249-262.

Persson L, Oredsson S, Anehus S, Heby O (1985) Ornithine decarboxylase inhibitors increase the cellular content of the enzyme: Implications for translational regulation. *Biochem Biophys Res Commun* 131: 239-245.

Popovic R, Kyle D, Cohen A, Zalik S (1979) Stabilization of thylakoid membranes by spermine during stress-induced senescence of barley leaf discs. *Plant Physiol* 64: 721-726.

Premecz G, Ruzicska P, Oláh T, Guyás A, Nyitrai A, Palfi G, Farkas G (1977) Is the increase in ribonuclease level in isolated tobacco protoplasts due to osmotic stress? *Plant Sci Lett* 9: 195-200.

Rhodes C, Lowe K, Ruby K (1988) Plant regeneration from protoplasts isolated from embryogenic maize cell cultures. *Bio/Technology* 6: 56-60.

Roberts DR, Dumbroff EB, Thompson JE (1986) Exogenous polyamines alter membrane fluidity in bean leaves - a basis for potential misinterpretation of their true physiological role. *Planta* 167: 395-401.

Santos M, Torné JM, Blanco J (1984) Methods of obtaining maize totipotent tissues. I. Seedling segments culture. *Plant Sci Lett* 33: 309-315.

Schwartz M, Altman A, Cohen Y, Arzee T (1986) Localization of ornithine decarboxylase and changes in polyamine content in root meristems of *Zea mays*. *Physiol Plant* 67: 485-492.

Slocum RD, Kaur-Sawhney R, Galston AW (1984) The physiology and biochemistry of polyamines in plants. *Arch Biochem Biophys* 235: 283-303.

Srivastava S, Vashi D, Naik B (1983) Control of senescence by polyamines and guanidines in young and mature barley leaves. *Phytochemistry* 22: 2115-2154.

Strogonov B, Shevyakova N, Kabanov Y (1972) Diamines in plant metabolism under conditions of salinization. *Fiziol Rast* 19: 1098-1104.

Tabor CW (1960) The stabilizing effect of spermine and related amines on mitochondria and protoplasts. *Biochem Biophys Res Commun* 2: 117-120.

Tabor H, Tabor CW (1964) Spermidine, spermine, and related amines. *Pharmacol Rev* 16: 245-300.

Tabor CW, Tabor H (1984) Polyamines. *Ann Rev Biochem* 53: 749-790.

Tiburcio AF, Masdéu MA, Dumortier FM, Galston AW (1986a) Polyamine metabolism and osmotic stress. I. Relation to protoplast viability. *Plant Physiol* 82: 369-374.

Tiburcio AF, Kaur-Sawhney R, Galston AW (1986b) Polyamine metabolism and osmotic stress II Improvement of oat protoplasts by an inhibitor of arginine decarboxylase. *Plant Physiol* 82: 375-378.

Tiburcio AF, Kaur-Sawhney R, Galston AW (1986c) Spermidine synthase as affected by osmotic stress. *Plant Physiol* 80: S-303.

Tiburcio AF, Kaur-Sawhney R, Galston AW (1987) Regulation by polyamines of plant tissue culture development *In:* Advances in the Chemical Manipulation of Plant Tissue Culture, Jackson MB, Mantell SH, Blake J (eds), Monograph No 16, British Plant Growth Regulator Group, pp 43-56.

Tiburcio AF, Kaur-Sawhney R, Galston AW (1988) Polyamine biosynthesis during vegetative and floral bud differentiation in thin layer tobacco tissue cultures. *Plant Cell Physiol* 29: 1241-1249.

Tiburcio AF, Gendy C, Tran Thanh Van K (1989) Morphogenesis in tobacco subepidermal cells: putrescine as marker of root differentiation. *Plant Cell Tiss Org Cult* 19: 43-54.

Tiburcio AF, Kaur-Sawhney R, Galston AW (1990a) Polyamine metabolism *In:* The Biochemistry of Plants, Intermediary Nitrogen Metabolism, Miflin BJ, Lea PJ (eds), Academic Press, New York, pp 283-325.

Tiburcio AF, Figueras X, Claparols I, Santos M, Torné JM (1990b) Polyamines and aging: effect of polyamine biosynthetic inhibitors on plant regeneration in maize callus cultured *in vitro In:* Plant Aging: Basic and Applied Approaches, Rodriguez R, Tames RS, Durzan DJ (eds), NATO ASI Series, Plenum Press, New York, pp 277-284.

Tiburcio AF, Figueras X, Claparols I, Santos M, Torné JM (1991) Improved plant regeneration in maize callus cultures after pretreatment with DL-alpha-difluoromethylarginine. *Plant Cell Tiss Org Cult* 27: 27-32.

Torné JM, Santos M (1990) The meristematic calluses of maize: a maintenance system of tissue juvenility *In:* Rodriguez R, Tames RS, Durzan DJ (eds), NATO ASI Series, Plenum Press, New York, pp 395-398.

Torrigiani P, Altamura M, Pasqua G, Monacelli B, Serafini-Fracassini D, Bagni N (1987) Free and conjugated polyamines during *de novo* floral and vegetative bud formation in thin cell layers of tobacco. *Physiol Plant* 70: 453-460.

Vasil V, Vasil I, Lu C (1984) Somatic embryogenesis in long-term callus cultures of *Zea mays* L (*Gramineae*). *Amer J Bot* 71: 158-161.

Vasil I (1987) Developing cell and tissue culture systems for the improvement of cereal and grass crops. *J Plant Physiol* 128: 193-218.

Winer L, Apelbaum A (1986) Involvement of polyamines in the development and ripening of avocado fruits. *J Plant Physiol* 126: 223-233.

Young D, Srinivasan P (1972) Regulation of macromolecular synthesis by putrescine in a conditional *Escherichia coli* putrescine auxotroph. *J Bacteriol* 112: 30-39.

The Ser/Thr protein kinase $p34^{cdc2}$ plays a central role in the regulation of the eukaryotic cell cycle (Nurse, 1990). Genetic research in *Sch. pombe* has shown that the *cdc2* gene product is required for both the start of the mitosis (M phase) as for the induction of DNA synthesis (S phase). A similar function has been attributed to the *CDC28* gene product, the $p34^{cdc2}$ homolog of *S. cerevisiae*. *cdc2*-Homologous genes have been amongst others, isolated from human, mouse, chicken, *Drosophila*, bean, alfalfa, maize, and *Arabidopsis thaliana* (Jacobs, 1992). In higher eukaryotes, the importance of $p34^{cdc2}$ in the G2/M transition is well known (Nurse, 1990), but contrary to yeast, there is most probably a $p34^{cdc2}$-like protein involved in the START transition (control point at the G1/S boundary of the cell cycle) (Pines and Hunter, 1990).

Studies in chicken cells showed that the regulation of mRNA levels is the most important factor which determines the amount of $p34^{cdc2}$protein during development. There is a clear correlation between the amount of *cdc2* mRNA and the proliferation of a tissue. However, in the analysis of individual embryogenic tissues, no correlation was found between the amounts of mRNA and protein, suggesting that post-transcriptional mechanisms contribute to control of $p34^{cdc2}$ protein levels (Krek and Nigg, 1989).

The activity of the $p34^{cdc2}$ kinase is regulated in a complex manner. $p34^{cdc2}$ can be phosphorylated on four different residues, depending on the cell type (Jacobs, 1992). Phosphorylation on each amino acid residue fluctuates during the cell cycle, suggesting that each has a potential regulatory role.

In fission yeast, *Sch. pombe*, $p34^{cdc2}$ is only phosphorylated on two sites during a normal cell cycle. One of these sites is a tyrosine residue on amino acid position 15 (Tyr_{15}) (Gould and Nurse, 1989). Tyr_{15} is conserved in all known $p34^{cdc2}$ proteins and has also been identified as a phosphorylation site in chicken and human $p34^{cdc2}$ homologs (Krek and Nigg, 1991; Norbury *et al.*, 1991). A phosphor on this residue, which is located in the ATP-binding domain, prevents ATP binding and thereby inhibits kinase activity. In all the above mentioned organisms (yeast, chicken, human), Tyr_{15} phosphorylation is restricted to the G2 phase of the cell cycle and during G2/M transition a dramatic dephosphorylation takes place, correlating with a high $p34^{cdc2}$ kinase activity (Gould and Nurse, 1989). Mutation of Tyr_{15} to Phe_{15} causes in *Sch. pombe* a premature mitosis (the Wee phenotype), suggesting that Tyr_{15} phosphorylation/dephosphorylation is crusial for the timing of $p34^{cdc2}$ activation and that of entry in mitosis.

The mitotic activator $p80^{cdc25}$ regulates directly Tyr_{15} dephosphorylation. Overexpression of $p80^{cdc25}$ causes a reduction in Tyr_{15} phosphorylation resulting in a premature start of the M phase (Russell and Nurse, 1986).

In higher eukaryotes, Thr_{14} has also been identified as a major phosphorylation site (Krek and Nigg, 1991; Norbury *et al.*, 1991). Phosphorylation of Thr_{14} works in conjunction with Tyr_{15} phosphorylation, inhibiting $p34^{cdc2}$ activation. In the presence of vanadate, which blocks Tyr dephosphorylation, $p80^{cdc25}$ can still partially dephosphorylate $p34^{cdc2}$, suggesting that $p80^{cdc25}$ can also dephosphorylate $p34^{cdc2}$ on Thr_{14} (Gautier *et al.*, 1991).

The $p34^{cdc2}$ protein is a key component of the mitose-promoting factor (MPF), which controls mitosis directly or indirectly (Jacobs, 1992). Activation of MPF drives the cell into mitosis, whereas inactivation leads to a return to interphase.

Activation of MPF does not only require the above mentioned activation of $p34^{cdc2}$, but also depends on the association of $p34^{cdc2}$ with another component of the MPF, cyclin

(Jacobs, 1992). It is the cyclin subunit which explains the periodic appearance of the MPF before M phase and its immediate disappearance after mitosis. Newly translated cyclin gradually accumulates during the S and G2 phases and disappear immediately during the metaphase/anaphase transition.

Cyclins possess characteristic sequences such as the cyclin box. Cyclins have now been cloned from a wide variety of organisms. On the basis of their sequences, they are divided into different classes such as A-cyclins, B-cyclins, and CLN-cyclins (Lew and Reed, 1992). The cyclin B kinase complex becomes activated just before mitosis, whereas A-cyclins bind $p34^{cdc2}$ earlier in the cell cycle, namely in S and G2. CLN cyclins function in the G1 phase of the cell cycle in *S. cerevisiae.*

CELL CYCLE GENES OF *ARABIDOPSIS THALIANA*

The molecular study of cell cycle regulation in plants is compared to yeast and animals, only in its initial stage. The first *cdc2* homolog of plants has been isolated by "polymerase chain reaction" in bean (Feiler and Jacobs, 1990). Since then, several other *cdc2* homologs have been isolated, more specifically in maize (Colasanti *et al.*, 1991), alfalfa (Hirt *et al.*, 1991), and *Arabidopsis* (Ferreira *et al.*, 1991; Hirayama *et al.*, 1991). The maize, alfalfa, bean, and one of the two *cdc2* genes from *Arabidopsis* (*cdc2a*) were shown to functionally complement a temperature-sensitive *cdc2* gene of *Sch. pombe.* We have shown that the *Arabidopsis cdc2a* gene product also functionally complements a temperature-sensitive *CDC28* mutant at non-permissive temperatures (Ferreira *et al.*, 1991, Fig. 1). The 1.4-kb *cdc2* cDNA insert was cloned into the yeast expression vector pEMBLyex4 downstream from the GAL-CYC1 hybrid promoter, yielding the plasmid pYCDC2Ara. This plasmid was introduced into the *S. cerevisiae* strain 4078-21a, which is unable to grow at 36°C because it carries a temperature-sensitive mutation in the *CDC28* gene. As can be seen in Figure 1, the yeast strain carrying pYCDC2Ara grows only at 36°C in the presence of galactose as inducer. Whereas the mutant strain does not grow at 36°C, cells carrying the pYCDC2Ara plasmid can form colonies at restrictive temperatures. However, fivefold to sixfold more colonies are formed in plates of wild-type cells and these colonies are clearly larger. This complementation is thus only partial which is not surprising given the fact that the *cdc2* gene product has to interact with numerous components of the cell cycle apparatus (e.g. cyclins, $p107^{wee1}$, $p80^{cdc25}$, and many substrates). Similar observations have been made with the human and *Drosophila cdc2* gene products and species-specific interactions could be responsible for the partial restoration of cell division (Lehner and O'Farrell, 1990). Alternatively, it remains possible that the partial complementation implies that the *Arabidopsis cdc2* gene product is only functional during G2/M transition. The three residues (Thr_{14}, Tyr_{15}, and Thr_{161}) of which the phosphorylation/dephosphorylation regulates the $p34^{cdc2}$ during G2/M transitions are completely conserved in the *Arabidopsis* protein. However, the $p34^{cdc2}$ Ser_{277} of which the phosphorylation peaks during the G1 phase and disappear early in the S phase, is substituted by an Asn_{277} in the *Arabidopsis* sequence. Possibly, the Ser_{277} phosphorylation/dephosphorylation of the $p34^{cdc2}$ is essential for G1/S transition. A second *cdc2* gene (*cdc2b*) in the *Arabidopsis* genome contains the Ser_{277} residue and support this hypothesis (Hirayama *et al.*, 1991).

A

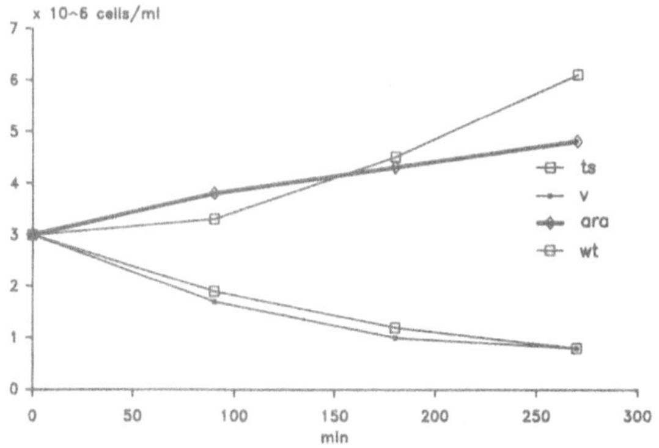

B

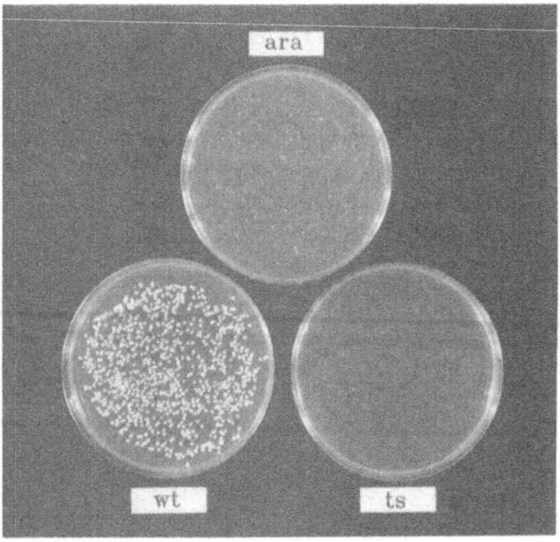

Figure 1. Functional complementation of a *S. cerevisiae CDC28* mutant strain by the *cdc2a* gene of *Arabidopsis thaliana*. A. Growth curve. Samples from liquid cultures containing 2% galactose growing at 36°C were taken every 90 min, plated on a rich medium, and incubated at 28°C. wt, *S. cerevisiae* strain 02431a; ts, *S. cerevisiae* strain 4078-21a carrying a temperature-sensitive *CDC28* gene (Hereford and Hartwell, 1974); ara, *S. cerevisiae* strain 4078-21a transformed with pYCDC2Ara; and v, *S. cerevisiae* strain 4078-21a transformed with pEMBLyex4. B. Aliquots taken after 1 h incubation at 36°C in a galactose-containing liquid medium, plated on a rich medium, and incubated for 3 days at 36°C. ara, wt, and ts, as in (A).

Recently, several cDNAs encoding cyclins were isolated from carrot, soybean (Hata *et al.*, 1991), and *Arabidopsis* (Hemerly *et al.*, 1992). The soybean cyclins are more like the B-type than the A-type cyclins and the carrot cyclin appears to be more like the A-type. However, the structural homology is not sufficient to classify these cyclins with certainty.

The *Arabidopsis* cyclin (Cyc1At) shows, inside the cyclin box, extensive homology with other representatives of the cyclin family, in particular mitotic cyclins (Hemerly *et al.*, 1992). The Cyc1At protein contains a mitotic destruction motif near the amino terminus of the protein and lacked the Pro-Glu-Ser-Thr (PEST) sequences near the carboxyl terminus, typical of the CLNs. However, it is again less clear whether it is an A- or B-type cyclin.

Figure 2. Amino acid sequence comparison of residues in the cyclin box. The amino acid sequence of Cyc1At from *Arabidopsis* is aligned with *Xenopus* cyclin A1 (Minshull *et al.*, 1990), C13-1 cyclin from carrot (Hata *et al.*, 1991), S13-6 cyclin from soybean (Hata *et al.*, 1991), and *Xenopus* cyclin B2 (Minshull *et al.*, 1990). Identical matches to the Cyc1At sequence are shaded. Amino acids indicated with an asterisk in the *Xenopus* A1 sequence and with an open circle in the *Xenopus* B2 sequence represent conserved motifs in all A-type and B-type cyclins, respectively.

Within the cyclin box, the Cyc1At protein is 47.3% identical to the *Xenopus* cyclin B2 and 46.3% identical to the *Xenopus* cyclin A1 (Fig. 2). The conserved amino acid motifs characteristic of either B-type (LRR-SK and HT-AKYL) or A-type cyclins (LVEV-EEYK and Q-LR) are not completely present in the Cyc1At protein sequence. Within the cyclin box, Cyc1At protein has 43% similarity with the carrot cyclin and 67.6% with soybean cyclin. At the moment, five different cyclin genes have been isolated from *Arabidopsis thaliana* and their analysis is currently in progress.

That the isolated *Arabidopsis* cyc1.At cDNA encodes indeed an active cyclin was shown by microinjection experiments of *Xenopus* oocytes (Hemerly *et al.*, 1992). *In vitro*-synthesized Cyc1.At mRNA was microinjected into G2-arrested oocytes and assayed for germinal vesicle breakdown, absence of nuclear envelope and white-spot formation. The results clearly demonstrated that Cyt1.At mRNA triggers the G2/M phase transition in oocytes, again underlying the extraordinary evolutionary conservation of the cell cycle control mechanism.

EXPRESSION ANALYSIS OF CELL CYCLE GENES

In higher eukaryotes, the expression of the *cdc2* gene is, at least in part, transcriptionally regulated (Krek and Nigg, 1989). RNA gel blot hybridizations revealed the presence of variable amounts of *cdc2* transcripts in all *Arabidopsis* tissues examined (Ferreira *et al.*, 1991). To study the expression at the cellular level, *cdc2a* promoter/β-glucuronidase fusions were constructed and analyzed in transgenic *Arabidopsis* plants (Hemerly *et al.*, in preparation). Independently, Martinez *et al.* (1992) have analyzed *cdc2* gene expression by *in situ* hybridizations in *Arabidopsis* and in radish. In general, the expression data obtained with both methods are highly comparable suggesting that, as in other eukaryotes, the *cdc2* gene expression is under transcriptional control.

The analysis of the *cdc2a* promoter/β-glucuronidase (Pcdc2a-GUS) chimeric genes allowed us to obtain a complete picture of *cdc2* gene expression during development. Early after seed germination, *cdc2* gene expression was only found in the root apical meristem. Somewhat later in development, expression is also observed in the shoot apical meristem and the zone which separates the root from the hypocotyl. This difference in timing of the

Gautier J, Solomon MJ, Booher RN, Bazan JF, Kirschner MW (1991) cdc25 is a specific tyrosine phosphatase that directly activates p34^{cdc2}. *Cell* 67:197-211.

Gould KL Nurse P (1989) Tyrosine phosphorylation of the fission yeast cdc2^{+} protein kinase regulates entry into mitosis. *Nature* 342:39-45.

Hata S, Kouchi H, Suzuka I, Ishii T (1991) Isolation and characterization of cDNA clones for plant cyclins. *EMBO J* 10:2681-2688.

Hemerly A, Bergounioux C, Van Montagu M, Inzé D, Ferreira P (1992) Genes regulating the plant cell cycle: isolation of a mitotic-like cyclin from *Arabidopsis thaliana*. *Proc Natl Acad Sci USA* 89:3295-3299.

Hereford LM, Hartwell LH (1974) Sequential gene function in the initiation of *S. cerevisiae* DNA synthesis. *J Mol Biol* 84:445-461.

Hirayama T, Imajuku Y, Anai T, Matsui M, Oka A (1991) Identification of two cell-cycle-controlling *cdc2* gene homologs in *Arabidopsis thaliana*. *Gene* 105:159-165.

Hirt H, Pay A, Gyorgyey J, Bako L, Nemeth K, Bogre L, Schweyen RJ, Heberle-Bors E, Dudits D (1991) Complementation of a yeast cell cycle mutant by an alfalfa cDNA encoding a protein kinase homologous to p34^{cdc2}. *Proc Natl Acad Sci USA* 88:1636-1640.

Jacobs T (1992) Control of the cell cycle. *Dev Biol* 153:1-15.

Krek W, Nigg EA (1989) Structure and developmental expression of the chicken *CDC2* kinase. *EMBO J* 8:3071-3078.

Krek W, Nigg EA (1991) Mutations of p34^{cdc2} phosphorylation sites induce premature mitotic events in HeLa cells: evidence for a double block to p34^{cdc2} kinase activation in vertebrates. *EMBO J* 10:3331-3341.

Lehner CF, O'Farrell PH (1990) *Drosophila cdc2* homologs: a functional homolog is coexpressed with a cognate variant. *EMBO J* 9:3573-3581.

Lew DJ, Reed SI (1992) A proliferation of cyclins. *Trends Cell Biol* 2:77-81.

Martinez MC, Jörgensen J-E, Lawton MA, Lamb CJ, Doerner PW (1992) Spatial pattern of *cdc2* expression in relation to meristem activity and cell proliferation during plant development. *Proc Natl Acad Sci USA* 89:7360-7364.

Minshull J, Golsteyn R, Hill CS, Hunt T (1990) The A- and B-type cyclin associated *cdc2* kinases in *Xenopus* turn on and off at different times in the cell cycle. *EMBO J* 9:2865-2875.

Nitschke K, Fleig U, Schell J, Palme K (1992) Complementation of the cs dis2-11 cell cycle mutant of *Schizosaccharomyces pombe* by a protein phosphatase from *Arabidopsis thaliana*. *EMBO J* 11:1327-1333.

Norbury C, Blow J, Nurse P (1991) Regulatory phosphorylation of the p34^{cdc2} protein kinase in vertebrates. *EMBO J* 10:3321-3329.

Nurse P (1990) Universal control mechanism regulating the onset of M-phase. *Nature* 344:503-507.

Pines J, Hunter T (1990) p34^{cdc2}: the S and M kinase? *The New Biologist* 2:389-401.

Russell P, Nurse P (1986) *cdc*$^{25+}$ Functions as an inducer in the mitotic control of fission yeast. *Cell* 45:145-153.

DEVELOPMENTAL AND ABSCISIC ACID REGULATION OF GENE EXPRESSION DURING EMBRYOGENESIS

Polydefkis Hatzopoulos

Department of Agricultural Biology and Biotechnology
Agricultural University of Athens, Iera Odos 75, 118 55 Athens, GREECE

INTRODUCTION

The basic features of development are established during embryogenesis. During this short period of the life cycle of a multicellular organism, a precise spatial and temporal arrangement of cells directs the "normal" developmental process. This principle is true for most if not all animal systems. Even though a similar principle can be postulated for plants, a detour of plant formation via somatic embryogenesis or organogenesis may pose intriguing questions. However, there are basic differences underlining the above mentioned processes between animal and plant systems: cell mobility and totipotency.

The elucidation of plant embryogenesis is fundamental to the understanding of the developmental process. Adult development involves plant growth from the two meristems which are formed early in the life cycle, during the mid stages of embryogenesis. The apical-basal polarity of the embryo is further maintained and orients plant enlargement according to the direction of embryonic axis. The basic body plan is established during the early stages of plant development. During the later stages of plant embryogenesis, embryos undergo fundamental developmental and physiological changes enabling them to enter the stage of maturation while appropriate environmental conditions initiate germination.

The first division of the zygote defines the apical and the basal cells. Embryonic cell division continues in a rapid cleavage pattern producing the octant-stage embryo and further a radially symmetrical structure, the globular-stage embryo. During this period, cells divide synchronously while overall cell size decreases. Differential cell division and cell growth enable embryos to grow with polarity and involves axis formation resulting in a structure with bilateral symmetry: the heart-stage embryo. As the heart-stage embryo elongates the three tissue layers, the protoderm, the cortex and the procambium are developed while the shoot and root primordia are established. Then, the embryo enters the torpedo-stage. Even though, the developmental cues that control the transition from one stage to another are

Morphogenesis in Plants, Edited by K.A. Roubelakis-Angelakis
and K. Tran Thanh Van, Plenum Press, New York, 1993

unknown, the establishment of somatic embryogenesis, a process resembling the ontogenetic events of zygotic embryogenesis, may provide insights to this transition.

The onset of embryogenesis involves new protein synthesis. Some are the seed storage proteins while others involved in synthesis, accumulation and storage of lipids and carbohydrates (Goldberg *et al.*, 1989). Regulatory proteins as well as pattern formation gene products modulating gene expression should also be present (Guiltinan *et al.*, 1990; Mayer *et al.*, 1991). The expression of seed (embryonic)-specific genes is developmentally regulated since their mRNAs and the respective products are only present in embryonic plant tissues. As a consequence of this developmental regulation as well as of the abundance of the gene products, the structure, function and expression of seed-specific genes and their products from numerous species have been studied extensively (Goldberg *et al.*, 1989). Another class of plant genes involved in oil biosynthesis and storage are also present in embryonic tissues (Bowman-Vance and Huang, 1988; Hatzopoulos *et al.*, 1990). A third class of plant genes responsive to developmental and hormonal cues does not exhibit a strict embryonic expression. Under "normal" physiological conditions, these genes are expressed only during embryo development. However, under conditions of high hormonal concentration or environmental stress these genes can be activated in organs not "normally" expressing them. Late embryogeny abundant (Lea) genes are an example of this class and they are believed to play a role of "protection" during seed desiccation (Hughes and Galau, 1991). Lea genes are conserved, the deduced polypeptides are hydrophilic and share common sequences within their structural regions when compared with many different species (Dure *et al.*, 1989). Lea genes as well as genes from the other classes are regulated by the plant hormone abscisic acid (Hetherington and Quatrano, 1991; Skriver and Mundy, 1990; Thomas *et al.*, 1992).

Abscisic acid (ABA) is a plant hormone which has been implicated in the control of a wide range of physiological responses in higher plants (Addicott, 1983). ABA is synthesized from mevalonic acid through the carotenoid biosynthetic pathway (Parry and Horgan, 1991; Zeevaart and Creelman, 1988). Numerous results have demonstrated that the sensitivity of the competent tissue to ABA, triggers a wide range of physiological processes such as, stomatal closure, inhibition of precocious germination of mature embryos and tolerance to stress. It has been shown that ABA inhibits germination, while 12 h after germination, embryos were no longer competent to respond to this plant hormone (Schopfer *et al.*, 1979). ABA plays a role antagonistic to the plant hormone gibberellin (GA) in the germination of mature embryos. Furthermore at the transcriptional level, ABA inhibits while GA induces alpha-amylase gene expression in cereals, in a tissue specific manner (Lanahan *et al.*, 1992; Skriver *et al.*, 1991). ABA levels and tissue sensitivity in response to ABA are two important parameters. In the *Arabidopsis* ABA-deficient mutant, ABA levels were reduced below the level of detection, but embryos developed normally, possibly because the mutant tissue has an increased sensitivity to the ABA (Koorneef *et al.*, 1982). In the *Arabidopsis* ABA-insensitive mutant, in which the embryonic tissue has a decreased sensitivity to ABA, the levels of the plant hormone were sufficient to establish "normal" embryo development (Koorneef *et al.*, 1984). However, in *Arabidopsis* double mutants in which the ABA levels and sensitivity to ABA were reduced, embryos exhibited precocious germination, desiccation intolerance and lack of storage protein accumulation (Koorneef *et al.*, 1989).

Furthermore, two other ABA-insensitive mutants, in barley and in maize showed that another component should exist in the ABA transduction pathway: a tissue specific component. The barley mutated leaves had "normal" ABA concentration but showed a decrease in the sensitivity of the stomatal response to ABA. However, this mutant exhibited a wild-type ABA inhibition of precocious germination (Raskin and Ladyman, 1988). The mutation in maize showed a seven-fold decrease in sensitivity to the inhibition of precocious germination by ABA (Robichaud *et al.*, 1980). However, the leaf tissues of this mutant showed a wild-type stomatal response to ABA. These studies clearly demonstrate that three parameters should be involved for "normal" physiological responses to this plant hormone: ABA level, tissue sensitivity to ABA and a tissue-specific ABA-mediated component. This does not necessarily account for three independent components present at all times in the embryonic tissues. The ABA level is influenced by its metabolism. The sensitivity as well as the tissue specific parameter can be integrated into a single component involved in the signal transduction pathway of the ABA response.

GENE EXPRESSION DURING EMBRYOGENESIS

A large number of genes is expressed during the stages of plant embryogenesis (Goldberg *et al.*, 1989). The same number of discrete mRNAs is present in almost all regions of the embryos as well as in endosperm tissues. However, this large number of genes expressed in plant embryos suggesting a high complexity at the molecular level, does not correlate with the view of a relatively simple structure of the embryo. From the anatomical point of view, the embryo consists of three types of tissues: the protoderm, the cortex and the procambium. In order to combine these two conflicting observations one or more of the listed speculations must be correct: i. Even though embryos superficially consist of three types of tissues, organized patches of cells from these tissues fate towards the formation of a particular structure. ii. The complexity of the mRNAs may not be that high. iii. A large number of genes designated for pattern formation should be actively transcribed in "subregions" of the embryos. iv. Almost all physiological processes should be in progress within the embryo. v. Multigene families expressed during embryogenesis may result in high complexity.

Most seed-specific gene expression is temporally and spatially regulated during embryogenesis and seed formation. These genes encode a variety of proteins that are accumulated during embryogenesis and stored in the dry seed (Goldberg *et al.*, 1989). Some of these are the storage proteins that serve as a nutrient pool of carbon and nitrogen for the germinating seedling; the functions of other seed proteins are not well understood (Higgins, 1984). These genes have been studied extensively because the regulatory processes controlling plant gene expression during development could be resolved. Transcription begins during late embryogenesis and early maturation (Walling *et al.*, 1986) and seed storage protein mRNAs are undetectable in mature plant organs. Another set of genes actively regulated during embryogenesis and transcribed from the globular-stage embryo are the Lea genes. The designated function of this diverse and hydrophilic class of proteins is the

DC8		
	-135	CCACGTGGCA
	-114	CTACGTGTTT
	-82	GTACGTGTCA
DC59		
	-263	TAACGTGGAC
	-130	AAAGGTGGCA
	-121	AGA-GTGGCA
DC24		
	-506	AAACGTGTCA
	-151	CCACG-GACC
	-89	CCACGCGTCC
	-57	ACACGTGCCC

CACGTG	ABRE
CACGTGGC	G-box
GACGTGGC	HEX

Figure 1. Comparison of upstream sequences of DC8, DC59 or DC24 which have homology to the ABRE motif. Deletions are represented by "-" and introduced to optimize the alignment. The G-box and HEX motifs are also shown (for full details see text).

scription, while the counterpart of DC59 is the furthest (Fig.1). Whether the position of the ABRE consensus sequence is combined ultimately to the rapid event of ABA induction has to be resolved.

Fluridone is an effective inhibitor of carotenoid biosynthesis and ABA accumulation. The incorporation of this agent into the culture medium showed interesting results. Fluridone was added in cultures of mid-phase somatic embryos and further allowed for 2-3 hours incubation. During this phase DC8, DC59 or DC24 mRNA transcripts were about at half of their maximum accumulation. Interestingly, all mentioned gene transcripts were decreased to almost undetectable levels, in the presence of fluridone. The pattern of mRNA accumulation in somatic embryos in the presence of fluridone resembled the one observed from very young embryos. Similar results were observed when the somatic embryos were grown in the presence of fluridone for a few days. The effect of fluridone was not deleterious to somatic embryos as, after several subcultures were put into fresh medium supplemented with fluridone, a proliferation of globular stage embryos was obvious. In a typical experiment of somatic embryogenesis all stages, globular, heart and torpedo are observed at average percentages of 60%, 20% and 20%, respectively. However, cultures containing fluridone had 95% globular stage embryos. This result indicates that fluridone and ultimately the absence of ABA inhibits or arrests the development of the embryos without affecting the viability of the cells (Hatzopoulos *et al.*, 1990b). Moreover, ABA exerts its role at the mid-to-late globular staged embryo during embryogenesis. When fluridone and ABA were supplied, these transcripts accumulated in a pattern resembling the one observed when ABA alone was included. However, the accumulation was less.

One point is important to note, the transcripts of DC8, DC59 or DC24 genes were detectable at very low levels when fluridone or very low levels of endogenous ABA were in somatic embryos, indicating that gene transcription had been initiated. It is plausible to speculate that the plant hormone ABA also plays an important role in the stability of mRNA transcripts. There is a profound and rapid decline of DC8 and DC59 mRNA levels in the presence of fluridone in somatic embryos already expressing these transcripts at high levels. However, the decline of ABA level in the presence of fluridone during the mid phases of embryogenesis may trigger specific mRNA degradation.

These results indicate that ABA regulates gene expression at many different levels. It increases/ decreases the steady-state levels of DC8, DC59 or DC24 mRNAs modulating the transcription of the corresponding genes as well as the post-transcriptional accumulation of these transcripts (Fig. 2). Moreover, DC8 protein could not be detected by antibodies in somatic embryos grown in the presence of fluridone but normally present during the developmental stages of embryogenesis (Hatzopoulos *et al.*, 1990b). Similar results obtained from other plants showed that ABA controls gene expression at the transcriptional as well as at the post-transcriptional level (DeLisle and Crough, 1989; Williamson and Quatrano, 1988). Em protein from wheat, could not be detected in seedlings, however transcript accumulation was similar when compared with the ones of the immature seed, while in the later case, the protein was clearly detected (Berge *et al.*, 1989; Williamson *et al.*, 1985). The data mentioned above, that is ABA decrease (in the presence of fluridone) in the mid phases of somatic embryogenesis with a simultaneous decrease of transcript accumulation to almost undetectable levels, contradicts the results obtained using mature and "dry" zygotic embryos. In this later case DC8 and DC59 mRNAs are easily detected (Borkird *et al.*, 1988), while it is known that the ABA level is rather low.

One could postulate that i) During the last stages of seed development (during desiccation) there is a decrease in tissue sensitivity to ABA. ii) The time response to ABA is developmentally regulated such that ABA modulates the expression (accumulation/reduction) of these gene products only during specified times of development (temporal) while later on physiological changes enable mature-dry embryos to keep high levels of these transcripts because fluctuations in ABA content does not exert its physiological response (Fig.2).

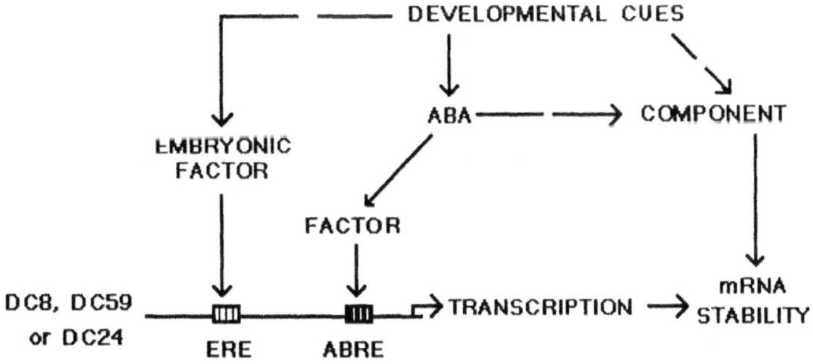

Figure 2. A model illustrating a possible transcriptional and post-transcriptional regulation of Lea and lipid body membrane protein gene expression during embryo development (for full details see text).

The intracellular levels of ABA content in mature tissues may be increased by desiccation (Zeevaart and Creelman, 1988) or by exogenously supplied ABA. DC8, DC59 or DC24 transcripts present only in embryonic tissues do not accumulate in vegetative tissues exposed to water stress by desiccation. Similarly, none of these transcripts is detected in mature tissues (leaves, petioles or roots) treated with high levels of ABA (Franz *et al.*, in preparation and unpublished data; Hatzopoulos *et al.*, 1990). However, when x-ray films were exposed for a long time to Northerns containing RNAs isolated either from desiccated or ABA treated tissues, a very faint but detectable signal was observed (unpublished data). Those signals many times had a degraded pattern, while in other cases they had a smaller molecular size. In all cases the signals were several orders of magnitude less (at least 1,000X less). However Em gene transcripts normally present in embryos accumulate in seedlings in response to ABA and desiccation (Berge *et al.*, 1989; Williamson *et al.*, 1985). Moreover, the transcripts of the 15 kD glycine-rich protein from maize normally appearing in the embryo, can be induced by ABA and desiccation in seedling and in leaves by wounding (Gomez *et al.*, 1988). The RAB 21 transcript from rice is present during embryogenesis but it is also induced in mature tissues when they are exposed to NaCl, desiccation or ABA (Mundy and Chua, 1988). The authors of the above mentioned results using Lea genes from monocots (i.e. RAB, Em and glycine-rich protein) investigated the modulation of ABA in young plant organs and seedlings, while the attempt to induce the DC8, DC59 and DC24 gene was in mature organs. Recently, it has been shown that RAB16B rice gene cannot be expressed in dicot leaves of tobacco transgenic plant (Yamaguchi-Shizonaki *et al.*, 1990). Moreover, B-19 like genes in barley and cotton are not induced in response to water-stress or ABA in non-embryonic tissues (Galau *et al.*, 1991). Interestingly, Espelund *et al.* (1992) have reported that the Lea B-19 gene family can be induced in immature embryos or imbibed mature embryos differentially by mannitol, NaCl or ABA. The later results indicate that there may be more than one transduction pathway involved in the modulation of expression of Lea genes in embryonic tissues by mannitol, NaCl or ABA.

To elucidate the mechanism that coordinates gene expression during embryo development, combined results obtained from gel retardation/DNase I footprinting analysis together with functional assays were evaluated. Gel retardation analysis showed that two adjacent 5' flanking sequences of the DC59 gene formed four complexes with nuclear extracts from embryos. Competition experiments revealed that the two adjacent DNA regions involved in the formation of the complexes do not compete with each other. It is unknown whether these *trans*-factors or *cis*-elements regulate the expression of DC59 gene but the complexes were detected only in nuclear extracts from embryos. Furthermore, the complexes formed between one upstream region of DC59 gene and nuclear extracts from embryos were competed with the 5' upstream region of the DC8 gene (Hatzopoulos *et al.*, 1990). DC8 and DC59 genes are coordinately regulated during embryo development. The same type of experiments showed that nuclear extracts from embryos contained DNA-binding proteins that interact *in vitro* with 5' flanking regions of DC8 gene (Goupil *et al.*, 1992). Interestingly, many DNA regions interacted with nuclear extracts from embryos in a specific way (unpublished data). However, two non adjacent regions of the 5' upstream sequences of DC8 gene (-155 to -505 and -658 to -1014) competed with each other. Moreover, the pattern of the gel-retarded bands was the same as the one observed from gel retardation experiments of

DC59 upstream region which competed with DC8 upstream regions. There is extensive homology between the upstream region of DC59 that competed with the corresponding of DC8, with the -155 to -505 fragment of DC8 gene. All three upstream regions mentioned have a high A/T content. Moreover, DNase I footprinting analysis showed that the protected sequences from DNase I digestion were within these homologous regions (Hatzopoulos *et al.*, 1990 and unpublished data). These upstream regions of the DC8 gene are necessary for gene expression (see later).

Partial purification of the nuclear proteins that bind to -155 to -505 region of DC8 gene showed that the polypeptides could be detected by Southwestern analysis and have medium molecular mass. This region was also used to screen a cDNA expression library made from mRNA isolated from somatic embryos. About five positive clones were detected and collected and are used for further analysis (unpublished data). There is one point that has to be mentioned: upstream regions of DC8 and DC59 genes forming complexes with nuclear proteins from somatic embryos did not comprise the ACGTG consensus ABRE sequences. Nevertheless, this motif is present within the upstream regions of DC8, DC59 or DC24 genes (Fig. 1).

It is reasonable to speculate that the *trans*-factors binding to -105 to -505 region of DC8 (or the ones binding to the DC59 upstream region) is the primary event *in vivo* while this complex formation modulates the ABA-mediated factor to interact with the ABRE sequences. An alternative explanation could be that the nuclear extracts contained insufficient amounts of the ABA-mediated *trans*-acting factor or that the functional property of this "regulatory" protein requires the continuous presence of ABA. During the biochemical isolation scheme, nuclear proteins were extracted usually from somatic embryos of carrots grown in medium without exogenously supplied ABA. The addition of ABA into the *in vitro* binding assays did not result in any additional complex when compared to the ones observed without any exogenously supplied ABA (unpublished data). However, Guiltinan *et al.* (1990) were able to detect complexes formed between ABRE sequences with nuclear extracts isolated from wheat embryos or rice suspension cultures imbibed or incubated, respectively in the presence of high levels of ABA (0.1 mM).

Functional analysis was developed by using transient expression assays in electroporated protoplasts. Gene constructs of various DC59 upstream regions linked to a reporter gene (beta-glucuronidase, GUS) were electroporated into carrot protoplasts. The transient assays showed a 2 to 4 fold induction of GUS activity after treatment with ABA (unpublished results). A more extensive analysis was followed for the upstream region of DC8 gene (Goupil *et al.*, 1992). Protoplasts were electroporated with various segments of the 5' flanking sequences of DC8, gene linked to GUS and 3' segment of NOS poly A^+ site. A promoter segment (-171 to -52) that contained the ABRE consensus sequences was able to induce almost two-fold GUS activity in transient assays following ABA treatment. However, when this segment was used in combination with the region -171 to -505 or with the -505 to -2600 region then a five-fold induction of GUS activity was observed in ABA treated protoplasts. As mentioned above, these two regions contain sequences interacting with nuclear extracts isolated from somatic embryos and competing with each other. Furthermore, it has been shown that the -171 to -505 segment contained long stretches of sequences with an extensive homology to the DNaseI protected upstream region of DC59.

ACKNOWLEDGMENTS

The author is grateful to Dr. R.Z. Sung for her inspiration and for providing the appropriate "intellectual and experimental environment". The author also wishes to thank Dr. G. Franz, Dr. P. Goupil, Mr. R. Chorneau and Mrs. D. Milioni.

REFERENCES

Addicott FT (1983) Abscisic acid. Praeger, New York.

Bartels D, Schneider K, Terstappen G, Piatkowski D, Salamini F (1990) Molecular cloning of abscisic acid-modulated genes which are induced during desiccation of the resurrection plant *Craterostigma plantagineum*. *Planta* 181: 27- 34.

Berge SK, Bartholomew DM, Quatrano RS (1989) Control of the expression of wheat embryo genes by abscisic acid. *In*: Molecular Basis of Plant Development, Goldberg RL (ed), Alan Liss, New York, pp 193-201.

Borkird C, Choi JH, Jin Z-H, Franz G, Hatzopoulos P, Chorneau R, Bonas U, Pelegri F, Sung ZR (1988) Developmental regulation of embryogenic genes in plants. *Proc Natl Acad Sci USA* 85: 6399-6403.

Bowman Vance V, Huang HCA (1988) Expression of lipid body protein gene during maize seed development. *J Biol Chem* 263: 1476-1481.

Bray EA, Beachy RN (1986) Regulation by ABA of beta-conglycinin expression in cultured developing soybean cotyledons. *Plant Physiol* 79: 746-750.

Bustos MM, Guiltinan MJ, Jordano J, Begum D, Kalman FA, Hall TC (1989) Regulation of beta-glucuronidase expression in transgenic tobacco plants by an A/T-rich, *cis*-acting sequence found upstream of a french bean beta-phaseolin gene. *The Plant Cell* 1: 839-853.

Choi JH, Liu L-S, Borkird C, Sung ZR (1987) Cloning of genes developmentally regulated during plant embryogenesis. *Proc Natl Acad Sci USA* 84 1906-1910.

Datta N, Cashmore AR (1989) Binding of a pea nuclear protein to promoters of certain photoregulated genes is modulated by phosphorylation. *The Plant Cell 1*: 1069-1077.

DeLisle AJ, Crouch ML (1989) Seed storage protein transcription and mRNA levels in *Brassica napus* during development and in response to exogenous abscisic acid. *Plant Physiol* 91: 617-623.

Dure III L, Crough M, Harada J, Ho T-HD, Mundy J, Quatrano R, Thomas T, Sung ZR (1989) Common amino acid sequence domains among the LEA proteins of higher plants. *Plant Mol Biol* 12: 475-486.

Espelund M, Saeboe-Larssen A, Hughes DW, Galau AG, Larsen F, Jakobsen SK (1992) Late embryogenesis-abundant genes encoding proteins with different numbers of hydrophillic repeats are regulated differentially by abscisic acid and osmotic stress. *The Plant J* 2: 241-252

Franz G, Hatzopoulos P, Jones JT, Krauss M, Sung ZR (1989) Molecular and genetic analysis of an embryonic gene, DC8, from *Daucus carota* L. *Mol Gen Genet* 218: 143-151.

Franz G, Hatzopoulos P, Fong F, Kalman K, Sung ZR The plant hormone abscisic acid enhances the expression of an embryonic gene, DC24, from *Daucus carota* L. in tissue specific manner, in preperation.

Galau GA, Hughes DW, Dure III L (1986) Abscisic acid induction of cloned cotton late embryogenesis abundant (Lea) mRNAs. *Plant Mol Biol* 7: 155-170.

Galau GA, Jakobsen KS, Hughes DW (1991) The controls of late dicot embryogenesis and early germination. *Physiol Plant* 81: 280-288.

Goldberg BR, Barker JS, Perez-Grau L (1989) Regulation of gene expression during plant embryogenesis. *Cell* 56: 149-160.

Gomez J, Sanchez-Martinez D, Stiefel V, Rigau J, Puigdomenech P, Pages M (1988) A gene induced by the plant hormone abscisic acid in response to water stress encodes a glycine-rich protein. *Nature* 334: 262-264.

Goupil P, Hatzopoulos P, Franz G, Hempel DF, You R, Sung ZR (1992) Transcriptional regulation of a seed specific carrot gene, DC8. *Plant Mol Biol* 18: 1049-1063.

Guiltinan MJ, Marcotte WR, Quatrano RS (1990) A plant leucine zipper protein that recognizes an abscisic acid response element. *Science* 250: 267-271.

Hajela K R, Horvath P D, Gilmour J S, Thomashow FM (1990) Molecular cloning and expression of cor (Cold-regulated) genes in *Arabidopsis thaliana*. *Plant Physiol* 93: 1246-1252.

Hatzopoulos P, Franz G, Choy L, Sung ZR (1990) Interaction of nuclear factors with upstream sequences of a lipid body membrane protein gene from carrot. *The Plant Cell* 2: 457-467.

Hatzopoulos P, Fong F, Sung ZR (1990) Abscisic acid regulation of DC8, a carrot embryonic gene. *Plant Physiol* 94: 690-695.

Hetherington MA, Quatrano SR (1991) Mechanisms of action of abscisic acid at the cellular level. *New Phytol* 119: 9-32.

Higgins TJV (1984) Synthesis and regulation of major proteins in seeds. *Annu Rev Plant Physiol* 35: 191-221.

Hughes DW, Galau GA (1991) Developmental and environmental induction of Lea and Lea A mRNAs and the postabscission program during embryo culture. *The Plant Cell* 3: 605-618.

Koorneef M, Jorna ML, Brinkhorst-Van Der Swan DLC, Karssen CM (1982) The isolation of abscisic acid (ABA) deficient mutants by selection of induced revertants in non-germination gibberellin sensitive lines of *Arabidopsis thaliana* (L) Heynh. *Theor Appl Genet* 61: 385-393.

Koorneef M, Reuling G, Karssen CM (1984) The isolation and characterization of abscisic acid-insensitive mutants of *Arabidopsis thaliana*. *Physiol Plant* 61: 377-383.

Koorneef M, Nanhart CJ, Hilhorst HWM, Karssen CM (1989) *In vivo* inhibition of seed development and reserve protein accumulation in recombinants of abscisic acid biosynthesis and responsiveness mutants in *Arabidopsis thaliana*. *Plant Physiol* 90: 463-469.

Lam E, Chua N-H (1991) Tetramer of a 21-base pair synthetic element confers seed expression and transcriptional enhancement in response to water stress and abscisic acid. *J Biol Chem* 266: 17131-17135.

Lanahan BM, Ho DT-H, Rogers WS, Rogers CJ (1992) A gibberillin response complex in cereal alpha-amylase gene promoters. *The Plant Cell* 4: 203-211.

Mansfield AM, Raikhel VN (1990) Abscisic acid enhances the transcription of wheat-germ agglutinin mRNA without altering its tissue-specific expression. *Planta* 180: 548-554.

Marcotte WR, Bayley CC, Quatrano RS (1988) Regulation of a wheat promoter by abscisic acid in rice protoplasts. *Nature* 335: 454-457.

Marcotte WR, Russel SH, Quatrano RS (1989) Abscisic acid-responsive sequences from the Em gene of wheat. *The Plant Cell* 1: 969-979.

Mayer U, Torres-Ruiz AR, Berleth T, Misera S, Jurgens G (1991) Mutations affecting body organization in the *Arabidopsis* embryo. *Nature* 353: 402-407.

Mundy J, Heigaard J, Hanson A, Hallgreen L, Jorgensen KG, Munck L (1986) Differential synthesis *in vitro* of barley aleurone and starchy endosperm proteins. *Plant Physiol* 81: 630-636.

Mundy J, Chua N-H (1988) Abscisic acid and water-stress induce the expression of a novel rice gene. *EMBO J* 7: 2279-2286.

Mundy J, Yamaguchi-Shizonaki K, Chua N-H (1990) Nuclear proteins bind conserved elements in the abscisic acid-responsive promoter of a rice Rab gene. *Proc Natl Acad Sci USA* 87: 1406-1410.

Napier JA, Chapman JM, Black M (1989) Calcium-dependent induction of novel proteins by abscisic acid in wheat aleurone tissue of different developmental stages. *Planta* 179: 156-164.

Parry AD, Horgan R (1991) Carotenoids and abscisic acid biosynthesis in higher plants. *Physiol Plant* 82: 320-326.

Pena-Cortes H, Sanchez-Serrano JJ, Willmitzer L, Prat S (1990) Abscisic acid is involved in the wound-induced expression of the proteinase inhibitor II gene in potato and tomato. *Proc Natl Acad Sci USA* 86: 9851-9855.

Pena-Cortes H, Willmitzer L, Sanchez-Serrano JJ (1991) Abscisic acid mediates wound induction but not developmental-specific expression of the proteinase inhibitor II gene family. *The Plant Cell* 3: 963-972.

Quatrano RS (1986) Regulation of gene expression by abscisic acid during angiosperm embryo development. *In*: Oxford Surveys of Plant Cell and Molecular Biology, Mifflin BJ (ed), vol 3, pp 467-477.

Quatrano RS (1987) The role of hormones during seed development. *In*: Plant hormones and their Role in Plant Growth and Development, Davies PJ (ed), Dordrecht/Kluwer Academic Publishing, pp 494-514

Raikhel NV, Wilkins TA (1987) Isolation and characterization of a cDNA clone encoding wheat germ agglutinin. *Proc Natl Acad Sci USA* 84:6745-6749.

Raskin I, Ladyman JAR (1988) Isolation and characterization of a barley mutant with abscisic acid insensitive stomata. *Planta* 173: 73-78.

Robichaud CS, Wong J, Sussex IM (1980) Control of *in vitro* growth of viviparous embryo mutants of maize by abscisic acid. *Devel Genet* 1: 325-330.

Schopfer P, Bajrachyra D, Plachy C (1979) Control of seed germination by abscisic acid I. Time course of action in *Sinapis alba* L. *Plant Physiol* 64, 821-827.

Schulze-Lefert P, Dangl JL, Beker-Andre M, Hahlbrock K, Schulz W (1989) Inducible *in vivo* DNA footprints define sequences necessary for UV light activation of the parsley chalcone synthase gene. *EMBO J* 8: 651-656.

Skriver K, Mundy J (1990) Gene expression in response to abscisic acid and osmotic stress. *The Plant Cell* 2: 503-512.

Skriver K, Lok-Olsen F, Rogers CJ, Mundy J (1991) *Cis*-acting DNA elements responsive to gibberellin and its antagonist abscisic acid. *Proc Natl Acad Sci USA* 88: 7266-7270.

Tabata T, Takase H, Takayama S, Mikami K, Nakatsuka A, Kawata T, Nakayama T, Iwabuchi M (1989) A protein that binds to a *cis*-acting element of wheat histone genes has a leucine zipper motif. *Science* 245: 965-967.

Thomas LT, Vivekananda J, Bogue MA (1992) ABA regulation of gene expression in embryos and mature plants. *In*: Abscisic acid: physiology and biochemistry, Davies WJ, Jones HG (eds), Bios Scientific Publications, pp 125-135.

Vinson CR, Sigler PB, McKnight SL (1989) Scissors-grip model for DNA recognition by a family of leucine zipper proteins. *Science* 246: 911-916.

Walling L, Drews GN, Goldberg RB (1986) Transcriptional and post-transcriptional regulation of soybean seed protein mRNA levels. *Proc Natl Acad Sci USA* 83: 2123-2127.

Williamson JD, Quatrano RS, Cumming AC (1985) Em polypeptide and its messenger RNA levels are modulated by abscisic acid during embryogenesis in wheat. *Europ J Biochem* 152: 501-507.

Williamson JD, Quatrano RS (1988) ABA-regulation of two classes of embryo-specific sequences in mature wheat embryos. *Plant Physiol* 86: 208-215.

Yamaguchi-Shizonaki K, Mino M, Mundy J, Chua N-H (1990) Analysis of an ABA-responsive rice gene promoter in transgenic tobacco. *Plant Mol Biol* 15: 905-912.

Zeevaart JAD, Creelman RA (1988) Metabolism and physiology of abscisic acid. Ann. Rev. Plant Physiol. *Plant Mol Biol* 39: 439-473.

OVERVIEW OF SHOOT MERISTEM FUNCTION DURING ANGIOSPERM DEVELOPMENT

During vegetative growth, the shoot apical meristems of higher plants appear to contain cells that function in a similar manner to the animal stem cells that are responsible for the continued growth of tissues such as epidermis and blood cells. Stem cells are distinguished by their ability to replenish themselves while also producing progeny that differentiate into specialized cell types (Siminovitch *et al.*, 1963). In plant meristems the apical initial cells appear to maintain some stem cell characteristics, and thus play a crucial role in plant growth. The inflorescence phase represents an important transition in normal plant development, for floral meristems that arise from the inflorescence meristem represent terminal cell populations.

There exist several lines of evidence for apical initial cells in the shoot apical meristem. Vegetative meristems often have an apical central zone comprised of enlarged, slowly dividing cells, and a flanking peripheral zone that contains smaller, more rapidly cycling cells. This zonation is thought to represent the restriction of function to different meristematic regions, with the central zone containing apical initial cells and their recent derivatives, and the peripheral zone encompassing the sites of organogenesis (see Steeves and Sussex, 1989; Wardlaw, 1965). Experimental evidence for the existence of apical initials comes from the analysis of clonal sectors in plants (Ruth *et al.*, 1985; Stewart and Derman, 1970). The occurrence of sectors that span large regions of the shoot, including the inflorescence, suggests that a small number of central zone meristematic cells serve as apical initials for shoot development, thought such initials are not necessarily permanent (Jegla and Sussex, 1989; Ruth *et al.*, 1985; Stewart and Derman, 1970). During normal meristem development, cells derived from these apical initials replenish the organogenic peripheral zone of the meristem where cells are recruited into developing organ primordia (see Steeves and Sussex, 1989).

The Shoot Apical Meristem During the Vegetative-to-Inflorescence Transition

The vegetative-to-inflorescence (V→I) transition, which marks the progression from vegetative to reproductive growth, is regulated by both inhibitory and promotive, graft-transmissible factors that are transported to the shoot tip (Evans, 1960; Lang *et al.*, 1977). It is well established that some of these substances are generated in leaf tissues (reviewed by Lang, 1965, and by Zeevaart, 1976), but other plant organs have been observed to influence the induction, initiation, and development of flowers (see Bernier *et al.*, 1986). The arrival of these substances at the shoot apex initiates inflorescence meristem development. The inflorescence meristem is the source of floral branches, leaf-like organs (e.g. bracts), and floral meristems, the latter of which produce the floral organs. This development results in the mature inflorescence, which is the collection of flowers produced by the plant that are often borne on floral branches.

In many plant species the V→I transition is associated with an increase in the mitotic activity of the central zone cells. The entire meristem becomes nearly uniform with regard to cell cycle times (Bodson, 1975; Corson, 1969; Steeves *et al.*, 1969), which eliminates the zonation observed in vegetative meristems. The transition is also marked by significant changes in cellular metabolism (Bernier, 1981). These changes in the organization and metabolic activity of the shoot apical meristem are likely to be important for establishing and maintaining the developmental potential of the shoot meristem during reproductive development.

The Shoot Apical Meristem During the Inflorescence-to-Floral Transition

In normal circumstances floral induction results in the commitment to flowering and the subsequent initiation of inflorescence and floral meristems (Hicks and Sussex, 1971). Inflorescence meristem function varies greatly among species. In plants that produce a determinate inflorescence, such as tobacco, this stage is short-lived and the shoot apical meristem is transformed into a terminal floral meristem. In plants producing an indeterminate inflorescence, such as *Antirrhinum* and *Arabidopsis*, the inflorescence meristem reestablishes a regenerative developmental strategy similar to that observed for the vegetative meristem. However, as inflorescence development progresses, the developmental potential of the apical meristem becomes restricted, and eventually the inflorescence terminates with the arrest of inflorescence meristem activity.

The inflorescence meristem is the source of floral meristems. Although anatomical differences between vegetative and floral organs are substantial, several lines of evidence suggest that leaves and floral organs are highly homologous. Morphological and evolutionary comparisons of the vegetative and floral structures of some plant groups have shown that floral organs are likely to be modified vegetative structures (Arber, 1937; Bailey and Swamy, 1951; Wilson and Just, 1939). Homology between floral and vegetative structures has also been suggested by nucleic acid hybridization analysis, which showed that a majority of mRNA species of tobacco reproductive structures (the anthers and ovaries) are common to the mRNA of tobacco leaves; in fact, the mRNA populations of petals and leaves are at least 98% homologous (Kamalay and Goldberg, 1980). Thus, the initial steps leading to the creation of a flower may result from the modification of the developmental program which operates in the vegetative shoot meristem during leaf initiation.

PROPER MERISTEM FUNCTION IS CRITICAL FOR NORMAL PLANT MORPHOGENESIS

It is likely that a limited repertoire of cellular processes (e.g. cell division and elongation) is used in the shoot apex during the life of the plant, but by a careful regulation of these processes a diverse set of organs can be initiated in a precise temporal and spatial sequence. Unfortunately, very little is known about how these cellular mechanisms are regulated in higher plants. If new genetic functions are important components of this regulation, then it is essential that we understand what these functions are, and how and when they are expressed. The first step in this process is to identify and isolate genes that facilitate the V→I→F progression. Once isolated, these genes can be used as markers for describing the cellular basis of meristem development. For example, it will be possible to determine which cells display coordinated responses to the V→I→F progression, and then to ascertain whether these cells are related by common developmental lineages or common cellular processes. In addition, it will be possible to begin to deduce the functions of these genes by using a number of molecular and biochemical techniques. As described below, the road to this goal has been long but the destination is in sight.

EVIDENCE FOR CHANGES IN GENE EXPRESSION DURING THE FLORAL TRANSITION

In the 1960's, a flurry of investigations focused on the involvement of both RNA and protein syntheses in the shoot apex during the vegetative-to-floral transition. Much of this work was concentrated on the cellular and biochemical events associated with floral evocation; evocation is defined as the commitment of the shoot apex to initiate floral development (Evans, 1969b; Knox and Evans, 1968). Numerous species were used in experiments designed to test the effect of metabolic inhibitors, analyze the incorporation of macromolecule precursors, and examine the patterns of histochemical stains. Later work assessed quantitative changes in individual proteins by immunological methods and 2-dimensional gel electrophoresis techniques. The majority of the rigorous investigations have been performed with single-cycle photoperiod sensitive varieties. These include the short day plants (SDP) *Pharbitis* nil and *Xanthium pennsylvanicum*, and the long day plants (LDP) *Lolium temulentum* and *Sinapis alba*. Only a few examples will be presented as several excellent reviews have been written on the biochemistry and physiology of flowering in individual genera (Evans, 1969a; Halevy, 1985).

Inhibitor Studies

Changes in the mRNA and protein complements within the shoot apical meristem upon evocation have long been thought to be intimately involved with, if not responsible for, inflorescence initiation. Some of the earliest studies designed to determine the nature of the meristematic biochemical processes associated with the vegetative-to-floral transition entailed the use of inhibitors, or antimetabolites, that prevented flowering. Such an approach has the potential to identify a specific metabolic process required for inflorescence initiation, and has particular utility in instances in which the inhibition is relieved by the simultaneous application of the corresponding metabolite.

Many factors warrant caution in the design and interpretation of antimetabolite experiments. The time course of inhibition and the transport to responding organs should be described. Secondary physiological effects should be mild or absent. Also, the site of action of the applied inhibitor should be identified, and this is particularly important when trying to discern between inductive processes within leaves and cotyledons versus evocational responses within apices. Where more than one biochemical pathway is affected, additional experiments are required to identify, if possible, which is responsible for the concomitant physiological consequence.

Despite these hurdles several elegant inhibitor studies provided persuasive evidence that newly synthesized gene products are essential for inflorescence and floral initiation. In order to assess the role of nucleic acid metabolism in the vegetative-to-floral transition, Salisbury and Bonner (1960) and Bonner and Zeevaart (1962) applied 5-fluorouracil (5-FU) and 5-fluorodeoxyuridine (5-FDU) to the leaves or buds of *Xanthium*. Their work led them to suggest that the biochemical action of 5-FU in floral inhibition was exerted upon nucleotide metabolism, possibly in nucleic acid synthesis. Evidence was provided that the process sensitive to 5-FU is renewed at the beginning of each dark period, regardless of the duration of the dark phase. Given that the floral stimulus was known to arrive at the *Xanthium* bud after the end of the dark period (Salisbury, 1969), one implication of this finding was that the V→I transition requires new nucleic acid synthesis in the bud. Alternatively, 5-FU could inhibit a process within the shoot apical meristem that responds to a fast-moving component of the leaf-generated stimulus (Evans, 1969b).

Given that 5-FDU was shown in other organisms to block the synthesis of thymidylic acid, 5-FDU was thought to be an inhibitor more specific to DNA synthesis (e.g. Cohen *et al.*, 1958). In fact, the inhibition of floral initiation in *Xanthium* by 5-FDU could be reversed by the application of thymidine, demonstrating that 5-FDU-specific floral inhibition was exerted likely through a disruption in DNA synthesis (Bonner and Zeevaart, 1962). It was noted, however, that such plants were also inhibited in vegetative growth; also, thymidine relieved the developmental inhibition when applied at the end, as well as at the beginning, of the inductive dark period. This suggested that the buds had generated and processed the necessary signals for induction and evocation, but due to a suspension of DNA synthesis (and probably, cell division) were unable to proceed with the developmental transformation. Once DNA synthesis was restored, the buds, having "remembered" the inductive processes, continued with inflorescence development. Thus, the authors proposed that, in *Xanthium*, the inhibition of the vegetative-to-floral transition caused by 5-FU and 5-FDU was likely to be a direct effect of blocking RNA synthesis.

In a similar series of experiments, Zeevaart (1962) examined the responses of *Pharbitis* to 5-FU and 5-FDU. Several important differences between the results obtained in these experiments versus those from the *Xanthium* experiments were seen. Notably, the inhibition of floral initiation caused by both 5-FU and 5-FDU was relieved by thymidine. A suppression in DNA synthesis was confirmed by incorporation experiments, and by the observation of mitotic figures in shoot tips. The evidence indicated that, in *Pharbitis*, DNA synthesis was required for floral evocation, and Zeevaart suggested that the genes responsible for floral differentiation were activated only during DNA replication.

Evans (1964) examined the effects of several inhibitors of nucleic acid and protein biosynthesis on the V→I transition in the LDP *Lolium temulentum*. Maximum inhibition of inflorescence initiation was achieved when 5-FU was applied at the end of the 24-hour long day, which coincides with the arrival of the leaf-generated floral stimulus (Evans and Wardlaw, 1964). Inflorescence development was markedly restored by orotic acid (a uridine monophosphate precursor), but not by thymidine. 5-FDU inhibited inflorescence initiation, but in contrast to the results obtained with 5-FU, the greatest inhibitory effect was seen two days following the end of the inductive long day. Kinet *et al.* (1971) also used 5-FU and 5-FDU to inhibit flowering in the LDP *Sinapis alba*. Both treatments showed similar inhibition kinetics to that described for *Lolium*. This time-dependent pattern suggested that DNA

that the activation of the rib meristem appears to be one of the earliest events in the vegetative-to-floral transition in *Pharbitis*. How this type of regional activation occurs is an important question that must be answered in order to understand the regulation of vegetative, inflorescence, and floral meristem function.

Biochemical Studies

The biochemical composition of the developing shoot apex has been under close scrutiny since the first chemical and electrophoretic tests were applied to apical extracts. The initial work followed general quantitative changes in protein content and nucleic acid chemistry during the floral transition, and several attempts were made to identify qualitative changes in protein and RNA populations. However, the techniques used were not sensitive enough to allow firm conclusions to be drawn. The progressive refinement in analytical techniques later enabled investigators to identify proteins specific to different developmental states of the shoot apex.

Perhaps the earliest attempt to identify changes in specific apical bud gene products during the V→I transition was conducted with *Sinapis*. Rabbit antisera were generated against protein extracts from vegetative buds, which possessed several leaf primordia, and reproductive buds that harbored numerous floral primordia, at various phases of development (Pierard *et al.*, 1977). Immunodiffusion tests were performed to identify proteins specific to buds from several developmental stages. However, it was later shown that these antigenic components identified by this screen were also present in vegetative organs.

Lyndon *et al.* (1983) applied 2-dimensional gel electrophoresis to extracts of *Sinapis* meristems as another approach to monitor changes in the protein complement during the V→I transition. Radioactively labeled methionine was topically applied to the apices of vegetative plants and to those which had been exposed to a 20 h inductive long day. The labeling and harvesting were conducted about 2 days following the beginning of the long day, and extracts were then subjected to electrophoresis. The authors noted about 14 proteins that increased in abundance following floral induction, and another 16 in the LD sample that appeared to be absent, or present in very low amounts, in the vegetative sample. Several proteins were observed to decrease upon floral induction. These results suggested that quantitative, and perhaps qualitative, changes in the protein composition of the shoot apex occur within a few days of floral induction. However, only a small number of proteins show a change in abundance with the vegetative-to-floral transition.

Ono *et al.* (1988) used 2-dimensional gel electrophoresis to separate *in vitro* translated products of bud RNA of *Pharbitis*. The comparison was between apices of plants exposed to 3SD cycles versus those from plants under 3 night-break (NB) interrupted SD. Three *in vitro* translated polypeptides were observed to be more abundant in the 3SD sample than in the NB sample. This reflects quantitative differences in specific mRNAs between induced and non-induced states.

Araki and Komeda (1990) also isolated proteins from *Pharbitis* apices following SD or NB treatment in another 2-dimensional gel electrophoresis study. Plants were subjected to a single SD, or SD with NB, and radioactive methionine was applied to apices 1, 4, 7, or 16 days later. Some apices also were labeled before the photoperiod. Quantitative changes in the levels of 5 proteins were noted: 2 steadily increased in SD conditions and remained low in NB exposures; one increased transiently regardless of light-regime; and two were more prominent under NB conditions.

Taylor *et al.* (1990) used the LDP *Silene coeli-rosa* in a similar study to determine if changes in meristem proteins were linked to the V→I transition or to cell synchronization. The inductive treatments were of 7LD followed by 2SD, or 7LD followed by 48 hours of darkness. The post-inductive 48-hour dark treatment disrupts synchronization of the cell cycle in the meristem and delays, but does not inhibit, the V→I transition. Non-inductive conditions were maintained as continuous SD. As compared with the SD controls, no change in the apical protein population was detected during the 7LD inductive treatment. However, over 50 additional proteins were observed at the end of both the 7LD + 2SD treatment and the 7LD + 48 hour dark treatment; also, the 2-dimensional patterns appeared identical for the two treatments, demonstrating that the changes in the observed meristem proteins were not dependent upon cell cycle synchrony in the *Silene* meristem.

The search to identify and isolate shoot apex proteins that either increase or decrease in abundance during the vegetative-to-floral transition has demonstrated that, while such proteins exist, the approach has limitations. The impediments are due largely to the difficulties in the isolation of a purified molecular clone of the DNA sequence that encodes the protein corresponding to a spot on a 2-dimensional gel. Without clones for the genes of interest, further investigations are nearly impossible due to the limited amount of shoot apex tissue, and thus proteins, that can be isolated for study.

IDENTIFICATION AND ISOLATION OF GENES ACTIVE IN THE SHOOT APEX DURING THE VEGETATIVE-TO-FLORAL TRANSITION

Improvements in contemporary biochemical techniques and the development of numerous methods to identify and clone genes only recently have enabled a pursuit of the genes that function in the shoot apex during the vegetative-to-floral transition. Such work had been hindered previously by the small amount of responsive tissue in the shoot apex. Progress in molecular biology, such as that seen in cDNA library construction and screening and in polymerase chain reaction (PCR) applications, and the exploitation of mutants and species possessing an abundance of floral meristem tissue, has led to the isolation of several genes with the tissue- and temporal-specific expression expected for genes involved in the V→I and I →F transitions.

One of the earliest attempts to use a direct cloning approach in order to isolate genes involved in the initiation of floral meristems was conducted with day-neutral tobacco by Meeks-Wagner *et al.* (1989). Due to the constraints imposed by the paucity of shoot apex tissue in mature plants, the thin cell layer (TCL) tissue culture system was employed to generate the necessary starting material for the cloning experiments (Tran Thanh Van, 1973). In the TCL system, floral branch explants raised on a kinetin-supplemented medium form floral meristems and eventually normal flowers, without intermediate callus formation or leaf production. Similarly, vegetative shoots grow from explants maintained on a zeatin-supplemented medium.

The cloning strategy entailed the construction of a cDNA library made from RNA of explants grown on the floral medium for 7 days, a stage at which active cell division occurs preceding meristem organization and organ differentiation. The library was differentially screened with cDNA probes made from RNA of day 7 floral and day 7 vegetative explants.

Five cDNA clones (FB7-1 to 5) were isolated that represented genes that produced transcripts abundant in floral explants but rare in vegetative explants. These genes were not specifically induced by kinetin, and thus the clones appeared to represent genes activated in association with *in vitro* floral initiation. Three clones represented low or single copy genes in the tobacco genome and were analyzed further. In whole plants the patterns of transcript accumulation were determined for these three FB7 clones by northern and *in situ* hybridization analyses. Transcripts were detected in several non-floral tissues and were most abundant in the roots. There appeared to be developmental regulation governing the accumulation of the FB7 transcripts in roots, in that the messages reached maximum steady-state levels shortly before the commitment of the shoot apical meristem to floral development. Based upon the well-documented influence of the leaves and roots upon floral initiation, the authors considered that FB7 gene expression in these organs could reflect an involvement in floral initiation.

It was later shown by sequence analysis that these genes encode a chitinase, a β-1,3-glucanase, and osmotin; also, transcripts of each gene were found to accumulate in response to physical stress (Neale *et al.*, 1990). It was suggested that the hydrolytic enzymes may be synthesized in response to changes in cell division patterns, or indirectly may activate genes involved in division and growth associated with the development of floral structures.

A subsequent screen of a TCL cDNA library made from explants cultured for 13 days on floral medium did not recognise genes that display transcriptional expression that is restricted to the shoot apex (G. Reid, C. Chang, and D.R.M-W., unpubl. results). This type of tissue specificity is anticipated for genes directly involved in the V→I and I→F transitions. Several groups have recently used this rationale as a basis for further attempts at cloning genes involved in the vegetative-to-floral transition.

Kelly *et al.* (1990) constructed a cDNA library from mRNA isolated from tobacco apices harvested during the vegetative-to-floral transition. Transition apices are devoid of floral organ primordia, but are taken from plants that are committed to floral development, as determined in rooting experiments similar to those of Singer and McDaniel (1986). The transition apex RNA was enriched for apex-specific transcripts by subtractive hybridization, and the resulting library was differentially screened with probes derived from vegetative apices and floral apices. Due to the limited amount of RNA extracted from the small quantities of available apical tissue, PCR amplification was used to generate the heterogeneous cDNA probes representing vegetative and floral apex mRNA populations.

Several clones were isolated that represented genes for which transcripts occur at a constant level in the shoot apex regardless of developmental stage, and genes for which increasing levels of transcripts occur in the shoot apex as the plant progresses through the V→I and I→F transitions. Clone FA2 displayed the greatest differential between vegetative and floral states and was further characterized. Northern analysis indicated that FA2 transcripts increased 50-fold in the shoot apex during the vegetative-to-floral transition. FA2 RNA was not detected in several leaf and root RNA samples. In addition, the transcripts were abundant in immature petals, stamens, and pistils, but less so in sepals. Sequence analysis of near full length clones identified no similarities in sequence data banks, nor common regulatory or structural peptide motifs (A.J.K. and D.R.M-W., unpubl. results). No predictions have been made regarding the function of the FA2 product, but given the temporal and tissue specificity, and the observation that FA2 transcripts accumulated in floral, but not vegetative, TCL explants, the evidence supports a role for the FA2 product in a general aspect of the V→I→F progression.

A similar cDNA cloning experiment was conducted by Melzer *et al.* (1990) with *Sinapis*. An advantage in the use of *Sinapis* is that the photoperiod requirement allows the synchronization of a large population of plants from which induced shoot apices can be harvested. Three cDNA libraries enriched for stage-specific clones by subtractive hybridization were constructed from RNA isolated from apices of plants exposed to 2, 5, and 10 inductive LD photoperiods. These libraries were screened with cDNA probes made from mRNA of the corresponding induced apices and cDNA probes from noninduced apices.

Two sets of clones were obtained. Six group I clones were isolated from the day 2 and day 5 libraries, and were characterized by the presence of transcripts in vegetative and induced apices, but not in leaves. The transcripts increased 5-to 10-fold in abundance following 2LD, continued to increase by 10LD, and then decreased as the floral buds matured. *In situ* hybridizations with clone pSFD2.43 indicated that the corresponding transcripts were abundant in the immature floral receptacle and present at lower levels in cells adjacent to the procambial strands. Clone pSFD5.04 hybridized to transcripts almost exclusively in the peripheral zone of the induced shoot apical meristem and in the flanks of young floral buds.

Group II clones isolated from the day 10 library represented transcripts that were not evident in apices until 5 to 10LD, reached maximum levels by 15LD, and then declined rapidly to undetectable levels in mature flowers. As with the group I clones, no transcripts were detected in several leaf samples. *In situ* hybridization demonstrated that all of the clones hybridized exclusively to RNA of the tapetum of young, unopened floral buds. The appearance and disappearance of group II RNAs likely coincided with pollen maturation followed by the dissolution of the tapetum.

The authors suggested that the group I clones do not represent genes that are common to all dividing cells. As demonstrated by the *in situ* analysis with pSFD5.04, specific transcripts accumulated in the peripheral zone of the floral apex, but not in the adjacent central zone, in which cells divide as rapidly as those in the peripheral zone upon reorganization of the meristem during the V→I transition. Thus, it was suggested that the corresponding genes of the group I cDNAs may be subject to a more compartmentalized function within the meristem, but most probably are not directly involved in floral organogenesis.

Differential cDNA cloning has been used recently to identify genes preferentially expressed in the shoot apex of *Lolium* (M. Zagotta, R. King, L. Evans, and D.R. M-W., unpubl. results). These genes produce unique mRNAs that are present in the shoot apex of mature vegetative plants. Transcripts of these genes were also detected in the *Lolium* shoot apex throughout the vegetative-to-floral transition, and during the initiation of floral organogenesis. While the putative proteins encoded by these genes have not been determined, Southern

both genes indicated that S2 and P4 RNA accumulated in numerous tissues, including the stem cortex, procambium, vascular cambium, and several floral and pod tissues. Transcripts were not detected in the shoot apical meristem or axillary meristems.

However, when a fragment of the S2 promoter spanning the sequences from -1200 to +45 was ligated to the GUS reporter gene and transformed into tobacco, the GUS staining patterns were markedly different from the transcript patterns revealed by *in situ* hybridization. Most notable was staining within the shoot apical and axillary bud meristems, and an absence of GUS protein in floral tissues. It was not determined whether the differences observed between the *in situ* and reporter gene results were due to inherent differences between pea and tobacco, or to the exclusion of some important regulatory regions in the cloning of the promoter. Since the promoters of several pea genes had been shown previously to confer identical expression patterns in pea and transgenic tobacco, the authors speculated that not all S2 regulatory elements had been included in the reporter gene construct. A deletion in the S2 promoter leaving a -115 to +32 fragment in the reporter fusion was shown to direct GUS expression strictly in the apical and axillary meristems of the shoot. A fragment of the P4 promoter extending from -248 to +32 similarly restricted expression to the shoot meristems.

A DNA footprint analysis of the deleted S2 promoter was conducted with extracts from cauliflower, and a repeated hexanucleotide sequence, termed the G-box, was bound by extract proteins. The G-box motif has been shown to be a required regulatory element for the expression of a diverse set of genes. Shoot meristem specificity was demonstrated in transgenic plants transformed with a chimeric reporter that consisted of a duplicated 25 base pair fragment containing the two G-boxes joined to a minimum heterologous promoter upstream of the GUS gene. The meristem expression was abolished in plants transformed with a similar construct containing a mutated version of the G-box. Thus, given the appropriate context, the minimum elements that confer meristem-specific expression have been identified for at least one gene; conceivably, the cis-acting sequences necessary for expression in floral tissue have also been delineated to the deleted -1200 to -115 region of the S2 promoter.

Atanassova *et al.* (1992) joined a 900 bp *Arabidopsis* histone H4 promoter fragment to the GUS gene to follow the tissue and developmental specificity of histone H4 expression in transgenic *Arabidopsis*. GUS staining was observed in cotyledons, shoot and root tips, leaves, stem, vascular cambium, and, with the exception of stamens, floral organs. A deletion of the sequences to position -219 upstream of the start of transcription reduced expression in all organs; however, transcription in non-meristematic tissues was more severely affected than in tissues with meristematic activity, such as the apical meristems, vascular cambium, and floral buds. A further deletion to -93 abolished nearly all H4-GUS expression. The authors thus suggested that cis-acting elements conferring meristematic specificity to the histone H4 promoter reside probably between nucleotides -93 and -219.

GENETIC STUDIES OF THE VEGETATIVE-TO-FLORAL TRANSITION

Recently the analysis of mutants that are defective in the vegetative-to-floral transition has played a major role in understanding shoot meristem function. Most of this work has made use of two plants: *Antirrhinum majus* and *Arabidopsis thaliana*. Mutants that have aberrant V→I and I→F transitions can be used for isolating the wild-type versions of the genes identified by the mutations, and, as described for the *anantha* mutant of tomato, as experimental systems for investigating gene expression during meristem development. At this time there are only a handful of mutations directly related to the early stages of the floral transition. Both the *centroradialis* mutant of *Antirrhinum* (Kuckuck and Schick, 1930) and the *terminal flower* 1 mutant of *Arabidopsis* (Alvarez *et al.*, 1992; Shannon and Meeks-Wagner, 1991) prevent normal inflorescence meristem development, and result in the production of a terminal flower. Thus, the wild-type *CENTRORADIALIS* and *TERMINAL FLOWER* 1 gene products appear to maintain the indeterminate potential of the inflorescence meristem by regulating a developmental pathway associated with the V→I transition.

The *floricaula*, *squamosa*, and *squamata mutants* of *Antirrhinum* (Coen, 1991; Coen and Meyerowitz, 1991; see Schwarz-Sommer *et al.*, 1990), and the *leafy*, *apetala* 1, and *apetala* 2 mutants of *Arabidopsis* (see Coen, 1991; Coen and Meyerowitz, 1991), result in floral meristems with characteristics of inflorescence meristems. Therefore, the wild-type alleles of these genes are required for the I→F transition.

The *FLORICAULA* (Coen *et al.*, 1990), *LEAFY* (Weigel *et al.*, 1992), and *SQUAMOSA* (Huijser *et al.*, 1992) genes have recently been cloned. DNA sequence analysis suggests that the gene products encoded by these genes are likely to be meristem-specific transcriptional regulators. This supports the hypothesis that the regulation of meristem gene activity is of major importance for proper plant development throughout the vegetative-to-floral transition. However, at present we have no indication as to the types of genetic functions controlled by these putative transcriptional regulators.

SUMMARY

Changes in gene expression in the shoot apical meristem have long been implicated in the switch from vegetative-to-floral development. Classical genetic approaches have identified some candidates involved in the early events of floral determination. For example, the recent analyses of the mutants of *Antirrhinum* and *Arabidopsis* have made significant contributions to our understanding of the genetic regulation of meristem function during the V→I and I→F transitions. However, many genes that contribute to the transition process may be missed in mutant screens due to silent, uninteresting, or lethal phenotypes. Indeed, it is known that some genes active in the floral meristems and floral organs are expressed in vegetative meristems before the plant is determined for reproductive growth (Kelly *et al.*, 1990; Shahar *et al.*, 1992); thus, mutant phenotypes may be expressed during vegetative development, and discounted as unrelated to the flowering process.

The goal of much of the work described in this review has been to identify markers with which to follow meristem function during plant morphogenesis. Recent advances in molecular techniques have enabled a complementary approach, the cloning of genes based upon tissue and temporal specificity, to map the sequence of events within the V→I→F progression. Some of the genes isolated in this manner may be products of a general metabolic enhancement that occurs during the cellular reorganization of the meristem during the vegetative-to-floral transition; these could include the histone genes (Alliotte *et al.*, 1989; Atanassova *et al.*, 1992; Kohler *et al.*, 1992; Koning *et al.*, 1991; Medford *et al.*, 1991), nucleic acid metabolism genes (Pri-Hadash *et al.*, 1992), and genes that encode putative ribosomal proteins (Kohler *et al.*, 1992).

As yet, only a relatively few genes have been cloned that appear to be involved in the vegetative-to-floral transition. However, available cDNA libraries that represent various tissues and developmental states in a number of species serve as useful experimental tools with which to pursue such genes. More sophisticated screens of existing libraries, or the PCR-based construction of libraries from tissue as specific as one or a few cells (Belyavsky *et al.*, 1989; Rappolee *et al.*, 1989) will aid in the identification of important molecular markers for analyzing early events in the shoot apex during the vegetative-to-floral transition. In the past three years, the first information describing genes expressed in the shoot apex during the V→I →F progression has been reported. We anticipate that a much more complete understanding of meristem function throughout plant development will be realized in the near future.

ACKNOWLEDGEMENTS

We thank Bryan Pickett, Susan Varnum, and Michelle Zagotta for their helpful comments on this manuscript. Also, we thank Dr. Mary E. Williams for communicating results prior to publication. The research from our laboratory reported in this review was supported by grants from NSF (DMB-8802163) and the McKnight Foundation. A.J.K. was supported by an NIH Developmental Biology Training Grant (5T32HD07348).

REFERENCES

Alliotte T, Tire C, Engler G, Peleman J, Caplan A, Van Montagu M, Inze D (1989) An auxin-regulated gene of *Arabidopsis thaliana* encodes a DNA-binding protein. *Plant Physiol* 89:743-752.

Alvarez J, Guli C, Yu X-H, Smyth D (1992) *Terminal flower*: a gene affecting inflorescence development in *Arabidopsis thaliana. Plant J* 2:103-116.

Araki T, Komeda Y (1990) Electrophoretic analysis of florally-evoked meristems of *Pharbitis nil* Choisy cv Violet. *Plant Cell Physiol* 31:137-144.

Arber A (1937) The interpretation of the flower: A study of some aspects of morphological thought. *Biol Rev* 12:157-184.

Arzee T, Zilberstein A, Gressel J (1975) Immediate intraplumular distribution of macromolecular syntheses following floral induction in *Pharbitis nil. Plant Cell Physiol* 16:505-511.

Atanassova R, Chaubet N, Gigot C (1992) A 126 bp fragment of a plant histone gene promoter confers preferential expression in meristems of transgenic *Arabidopsis. Plant J* 2:291-300.

Bailey IW, Swamy BGL (1951) The conduplicate carpel of dicotyledons and its initial trends of specialization. *Am J Bot* 38:373-381.

Belyavsky A, Vinogradova T, Rajewsky K (1989) PCR-based cDNA library construction: general cDNA libraries at the level of a few cells. *Nucl Acids Res* 17:2919-2932.

Bernier G (1986) The flowering process as an example of plastic development. *In:* Plasticity in Plants, Jennings EH, Trewavas AJ (eds), Company of Biologists, Cambridge, pp 257-286.

Bernier G (1988) The control of floral evocation and morphogenesis. *Annu Rev Plant Physiol Plant Mol Biol* 39:175-219.

Bernier G, Kinet J-M, Sachs RM (1981) The Physiology of Flowering, vol II, CRC Press, Boca Raton, Florida.

Bodson M (1975) Variation in the rate of cell division in the apical meristem of *Sinapis alba* during the transition to flowering. *Ann Bot* 39:547-554.

Bonner BA, Gifford EM, Reed N-MR (1991) Differential RNA synthesis in the shoot apex of *Pharbitis nil* during floral evocation. *Am J Bot* 78:401-407.

Bonner J, Zeevaart, JAD (1962) Ribonucleic acid synthesis in the bud an essential component of floral induction in *Xanthium. Plant Physiol* 37:43-49.

Bronchart R, Bernier G, Kinet J-M, Havelange A (1970) RNA synthesis in the cells of the apical meristem of *Sinapis alba* during transition from the vegetative to the reproductive condition. *Planta* 91:255-269.

Coen ES (1991) The role of homeotic genes in flower development and evolution. *Annu Rev Plant Physiol Plant Mol Biol* 42:241-292.

Coen ES, Meyerowitz EM (1991) War of the whorls: genetic interactions controlling flower development. *Nature* 353:31-37.

Coen E, Romero J, Doyle S, Elliot R, Murphy G, Carpenter R (1990) *Floricaula*: A homeotic gene required for flower development in *Antirrhinum majus. Cell* 63:1311-1322.

Cohen SS, Flaks JG, Barner HD, Loeb MR, Lichtenstein J (1958) The mode of action of 5-fluorouracil and its derivatives. *Proc Natl Acad Sci USA* 44:1004-1012.

Corson GE Jr (1969) Cell division studies of the shoot apex of *Datura stramonium* during the transition to flowering. *Am J Bot* 56:1127-1134.

Evans LT (1960) Inflorescence initiation in *Lolium temulentum* L. II. Evidence for inhibitory and promotive photoperiodic processes involving transmissible products. *Aust J Biol Sci* 13:429-440.

Evans LT (1964) Inflorescence initiation in *Lolium temulentum* L. VI. Effects of some inhibitors of nucleic acid, protein, and steroid biosynthesis. *Aust J Biol Sci* 17:24- 35.

Evans LT (1969a) The Induction of Flowering. Macmillan, Melbourne.

Evans LT (1969b) The nature of flower induction. *In:* The Induction of Flowering, Evans LT (ed), Macmillan, Melbourne, pp 457-480.

Evans LT (1969c) *Lolium temulentum* L. *In:* The Induction of Flowering, Evans LT (ed), Macmillan, Melbourne, pp 328-349.

Evans LT, Wardlaw IF (1964) Inflorescence initiation in *Lolium temulentum* L. IV. Translocation of the floral stimulus in relation to that of assimilates. *Aust J Biol Sci* 17:1-9.

Gifford EM Jr (1964) Developmental studies of vegetative and floral meristems. *Brookhaven Symp Biol* 16:126-136.

Halevy AH (1985) Handbook of Flowering. CRC Press, Boca Raton.

Hicks GS, Sussex IM (1971) Organ regeneration in sterile culture after median bisection of the flower primordia of *Nicotiana tabacum. Bot Gaz* 132:350-363.

Huijser P, Klein J, Lonnig W-E, Meijer H, Saedler H, Sommer H (1992) Bracteomania, an inflorescence anomaly, is caused by the loss of function of the MADS-box gene squamosa in *Antirrhinum majus. EMBO J* 11:1239-1249.

Jegla DE, Sussex IM (1989) Cell lineage patterns in the shoot meristem of the sunflower embryo in the dry seed. *Dev Biol* 131:215-235.

SECRETED PROTEINS IN PLANT CELL CULTURES

Fred A. van Engelen and **Sacco C. De Vries**

Department of Molecular Biology
Agricultural University Wageningen
Dreijenlaan 3, 6708 HA Wageningen, THE NETHERLANDS

INTRODUCTION

The possibility of *in vitro* culture of plant tissues and cells has provided a means of studying fundamental problems in plant biology. Two research areas especially have benefited from the use of plant cell suspension cultures. First, the analysis of plant primary cell walls has made considerable progress through the use of cell suspensions as a relatively homogeneous source of primary cell walls (Keegstra *et al.*, 1973). Second, the ability of cultured cells of carrot and many other species to form embryos that are almost identical to embryos derived from the zygote has been exploited by those studying plant embryo development. Major advantages of the experimental system of somatic embryogenesis, the term given to embryo development from cultured somatic cells, over zygotic embryogenesis are the very high numbers of embryos that can be obtained in a short time and the accessibility of the embryos for experimental manipulation. Although much work on somatic embryogenesis has been done in the three decades that have passed since its original demonstration (Reinert, 1959), there still are large gaps in our understanding of embryo development, especially with respect to the molecular mechanisms that regulate it (Dudits *et al.*, 1991).

Every plant cell is enveloped by a cell wall. Two generalized types of cell walls can be discerned. Primary cell walls are thin (about 100 nm) and are deposited by dividing and expanding cells. As growth ceases and cells acquire specialized functions, they may form a more rigid secondary wall. The plant primary cell wall now is believed to generally play an important part in the early phases of plant growth and development. This wall is faced with the complex task of providing a strong skeleton for the organism as a whole and protecting it from environmental influences, such as attack by pathogens, while at the same time it has to retain the plasticity to accommodate controlled cell extension in response to turgor pressure. The latter process is crucial for morphogenesis, since, due to the very presence of the wall, plants can only determine their shape by regulation of the rate and plane of cell division and

Morphogenesis in Plants, Edited by K.A. Roubelakis-Angelakis
and K. Tran Thanh Van, Plenum Press, New York, 1993

the degree and direction of expansion of the cells after division. To fulfil this task, higher plants have evolved an intricate wall structure composed of many different polymers. Regulation of wall properties can be achieved by secretion of new wall constituents or enzymes that modify the polymers that are already present. Indeed numerous activities of enzymes that have cell wall polymers as their potential substrates have been detected in the cell wall (Fry, 1985).

The growth medium of plant cell cultures may be regarded as a large extension of the intercellular space; soluble secreted molecules that inhabit the apoplast in planta will accumulate in the medium when cells are grown in suspension (Stevenson *et al.*, 1986). Hence, growth and development of cultured cells are likely to be reflected in the molecules that are secreted into the culture medium. This opens up the possibility of studying the role of these molecules in early plant development by analyzing the culture medium during somatic embryogenesis.

The lack of consensus on the definition of a plant cell (Sack, 1991; Staehelin, 1991; Stafford, 1991) may cause confusion about some of the terms used. With "extracellular" we refer to everything outside the plasma membrane, and the term is equivalent to "intercellular" when a group of cells is considered. The continuum of extracellular space(s) is called the apoplast. Thus, the cell wall is extracellular and, by analogy with animal cells, may be called "extracellular matrix" (Roberts, 1989).

This review will provide a summary of the constituents and assembly of the plant primary cell wall as well as the biochemical processes that are thought to take place in it. In this background, the possible functions of proteins secreted into plant cell culture media, with special focus on somatic embryogenesis, are discussed. Finally, some questions regarding the control of extracellular protein function and interactions between different cell types in embryogenic plant cell cultures are addressed.

COMPONENTS OF THE PRIMARY WALL

Any attempt to unravel the roles of the primary cell wall and the proteins that are associated with it, in terms of plant developmental processes, must set out with a description of the structure and properties of its constituents. The primary cell wall has a water content of 70 to 80%, and its solids are composed - on average - of 90% polysaccharides and 10% proteins (McNeil *et al.*, 1984). Both of these may be subdivided further into classes of polymers with widely differing properties. This complex array of cell wall polymers is described below. The focus will be on dicotyledonous (dicot) species with occasional reference to *Gramineae* (grasses, cereals), the best studied monocotyledonous (monocot) family.

Polysaccharides

Cell wall polysaccharides can be classified into three groups: pectins, hemicelluloses and cellulose. This classification has been based on procedures used sequentially to extract polysaccharides from the wall. The different components of the fractions obtained have been found to share certain properties and the classical terminology has been maintained. Because the literature on polysaccharides is massive, most of the following information has been extracted from a number of review articles (Darvill *et al.*, 1980, 1985; Fry, 1985, 1988,

1989a,b, 1990; Jarvis, 1984; Kato, 1981; McCann and Roberts, 1991; McNeil *et al.*, 1984; Roberts, 1989; Varner and Lin, 1989). For further details, we refer to the papers cited therein.

Pectins

Pectins are an extremely complex mixture of polysaccharides that make up 15-35% of the dicot primary cell wall (dry weight) and are the major constituents of the middle lamella, though the pectins of the latter are of a different composition. Most, but not all (Selvendran, 1985), pectic polysaccharides can be extracted with hot water, ammonium oxalate, weak acid or chelating agents. The majority of the polysaccharides obtained are rich in galacturonic acid, but some are composed largely of galactose and/or arabinose. Since the isolation procedures are likely to cleave covalent bonds (Selvendran, 1985), the different polymer types described below are thought currently to be domains of larger molecules *in muro* (Jarvis, 1984). Pectins then may be defined as galacturonic acid-rich polysaccharides, the side chains of such polysaccharides, and chemically similar polysaccharides (Fry, 1988). Graminaeceous monocots contain very low levels of pectin (~3%). Other monocots are closer to dicots or have intermediate amounts of pectin (Jarvis *et al.*, 1988).

Homogalacturonan. Homogalacturonan is an unbranched chain of α-(1-4)-linked galacturonic acid with occasional rhamnose residues that give rise to kinks in the molecule. At the time of secretion a large proportion of the carboxyl groups is methyl-esterified and these are distributed in blocks (Tuerana *et al.*, 1982). In some pectins hydroxyl groups of galacturonic acid residues are acetylated.

Rhamnogalacturonan I. Rhamnogalacturonan I (RGI) consists of 1000-2000 alternating residues of galacturonic acid and rhamnose. About 50% of the rhamnose units bear side chains of 1 to more than 15 arabinose and/or galactose residues in many (up to 30) different branched structures (McNeil *et al.*, 1984). In some *Chenopodiaceae* (spinach and beet) the phenolic groups ferulate and *p*-coumarate have been found to be attached to terminal arabinose or galactose (Fry, 1988).

Rhamnogalacturonan II. The structure of rhamnogalacturonan II (RGII) is quite different from RGI. In sycamore it is a small, highly branched molecule with only about 60 glycosyl residues, containing 12 different sugars (Fry, 1988) among which many exotic ones like 2-O-methylglucose, 2-O-methylxylose, apiose, 3-C-carboxy-5-deoxyxylose and 3-deoxy-manno-octulosonic acid. It has a galacturonic acid-rich backbone, with several copies of two different heptasaccharides and some smaller oligosaccharides as side chains. Because RGII is solubilized from the wall with an endopolygalacturonase it is believed to be covalently linked to wall polymers through homogalacturonan stretches.

Neutral pectins. This class comprises arabinans, galactans and arabinogalactans with a composition that is similar to the side chains of RGI. Because pectins are susceptible to degradation, it has been argued that these neutral polysaccharides are an artefact of the isolation procedure (Selvendran, 1985). However, some of the neutral pectins are much larger than the side chains of RGI, suggesting that they form a separate group.

Enzymes

The structural components of the cell wall are subject to the action of numerous enzymes, most of which appear to have a catabolic function. Some enzymes can diffuse freely in the apoplast, others are bound to the wall through ionic interactions or covalent bonds. Known cell wall enzyme activities have been listed by Fry (Fry, 1985, 1988). Therefore, only a few examples of each class of enzymes are given below.

Glycosidases. Glycosidases release monosaccharides from the non-reducing end of a polysaccharide. These enzymes usually have a high specificity for released sugar and its linkage type, but their activity is much less dependent on the polymer from which the sugar is cleaved. Examples are β-glucosidase, α-galacturonidase and α-mannosidase.

Endoglycanases. These enzymes catalyse the hydrolysis of carbohydrate linkages in mid-chain. Examples are endopolygalacturonase, cellulase and endo-β-1,3-glucanase. As for glycosidases, the cleavage sites of endoglycanases may occur in different kinds of substrates.

Transglycosylases. An enzyme is said to have transglycosylase activity if after cleavage it transfers the released (chain of) sugar residue(s) to a different chain, thereby conserving the total number of glycosidic bonds. An endo-trans-xyloglucanase activity recently has been demonstrated *in vivo* (Smith and Fry, 1991).

Other hydrolases. A variety of esterase, protease and phosphatase activities has been detected in the wall. Some esterases may act on pectins, cleaving methyl-, acetyl- or feruloyl-ester bonds. Proteases may degrade any of the wall proteins, but no specific cases have been described yet. Extracellular substrates for phosphatase are not known.

Peroxidases. Most peroxidases are heme-containing glycoproteins that can catalyse two kinds of reactions: $RH_2 + H_2O_2 \rightarrow R + 2H_2O$ or $2RH + H_2O_2 \rightarrow R_2 + 2H_2O$ in which R is usually an aromatic ring. The latter reaction results in intra- or intermolecular crosslinking as exemplified by the peroxidase-catalysed polymerisation of lignin (Catesson *et al.*, 1986) or the coupling of tyrosine residues of extensin (Everdeen *et al.*, 1988). A large number of different peroxidases are found that differ in their isoelectric point, (artificial) substrate preference and/or organ-specific presence (Greppin *et al.*, 1986).

Oxidases. Substrates for oxidases are ascorbate, polyphenols, indoleacetic acid and polyamines. The oxidation of polyamines generates H_2O_2, which has been inferred to be a mechanism to control the activity of peroxidases (Angelini and Federico, 1989).

Malate dehydrogenase. Malate dehydrogenase activity results in the formation of reduced NAD. The oxidation of reduced NAD by peroxidase results in the formation of H_2O_2. This cycling of NAD in the wall may be used as an alternative mechanism to generate the H_2O_2 required for peroxidase activity.

Other Proteins

Thionins. Cell walls of the endosperm and leaves of cereals contain high levels of thionins (Bohlmann *et al.*, 1988; Garcia-Olmedo *et al.*, 1989). These small polypeptides have an antimicrobial activity, but their mode of action is unknown.

Arabinogalactan proteins (AGPs). The AGPs are a family of proteoglycans which have in common that their carbohydrate content is very high (90-98%) and consists mainly of arabinose and galactose in varying branched structures (Fincher *et al.*, 1983). In addition, the protein core is rich in hydroxyproline and in serine, alanine and glycine. AGPs have a lectin-like activity (see below).

Lectins. A lectin has been defined broadly as a carbohydrate-binding protein other than an antibody or an enzyme (Barondes, 1988). Several lectins have been extracted from the cell walls of members of the *Leguminosae and Gramineae* (Etzler, 1985). It is not known whether similar extracellular lectins exist in other plant families. AGPs are able to bind to artificial carbohydrate antigens and may therefore also be regarded as lectins (Fincher *et al.*, 1983). However, binding to other cell wall components has not yet been reported.

CONSTRUCTION AND FUNCTIONING OF THE PRIMARY CELL WALL

In contrast to the structure of many of the cell wall's building blocks, knowledge of the assembly of these components into an ordered functional structure is very incomplete. The wall polymers probably are held together by a variety of intermolecular bridges, such as hydrogen, ionic or covalent bonds, thus giving rise to a viscoelastic gel. It is still unclear which of the cross-links are involved in the regulation of wall mechanics. In recent dicot wall models (Roberts, 1989; Talbott and Ray, 1992; Varner and Lin, 1989) the presence of three separate polymer systems has been proposed: a cellulose/xyloglucan network, an extensin network and a pectin network. These networks appear to be assembled into a functional wall structure through non-covalent interactions and physical entanglement (Talbott and Ray, 1992a,b).

Cellulose/Xyloglucan Network

There is growing evidence that a cellulose/xyloglucan network is a central factor in the regulation of cell expansion (Fry, 1989). NMR and binding studies (Hayashi *et al.*, 1987; MacKay *et al.*, 1988) have suggested that in the wall the cellulose microfibrils are coated with xyloglucan through hydrogen bonds. Because of their size and flexibility, individual xyloglucan molecules may bind to and thereby crosslink different microfibrils which results in a network. Electron microscopy studies support this configuration (McCann *et al.*, 1990). With such a network, expanding cells encounter resistance from xyloglucan molecules that have become stretched taut between different cellulose microfibrils. Continued expansion would then require release of hydrogen bonds or enzymatic cleavage of the stretched xyloglucan chains, because the cellulose microfibrils are not extensible. A model has been proposed for auxin-induced wall loosening in stem and coleoptile growth involving a cellulase that hydrolyzes xyloglucan rather than cellulose *in vivo* (Fry, 1989). According to this model, auxin increases the endo-xyloglucanase activity by *de novo* synthesis of cellulase or by proton secretion which activates the enzyme. Undesired weakening of the cell wall by xyloglucan cleavage may be prevented if the enzyme also has a trans-xyloglucanase activity that reforms bonds between different segments. Such an activity indeed has been found in expanding suspension cell walls (Smith and Fry, 1991).

Other enzymes that may augment the susceptibility of xyloglucan to enzymatic cleavage are glycosidases able to remove side chains from the xyloglucan backbone, viz. α-fucosidase, β-galactosidase and α-xylosidase.

In graminaeceous monocots the role of xyloglucan is probably fulfilled by mixed-linkage β-glucan.

Extensin Network

Extensin is secreted as a soluble protein but gradually becomes inextractable from the wall, which indicates that the protein is somehow coupled to the wall. That this coupling takes the form of a covalently crosslinked network has been inferred from several observations. First, treatment of complete cell walls with hydrogen fluoride dissolves all carbohydrates, but an insoluble structure consisting of crosslinked proteins, including extensin, remains (Mort and Lamport, 1977). Second, the tyrosine dimer isodityrosine has been found in the wall (Fry, 1982) and in purified tomato extensin, though at least some of the bridges in the latter are intramolecular (Epstein and Lamport, 1984). Third, covalently bound extensin can be solubilized by an agent that can specifically cleave isodityrosine (Biggs and Fry, 1990). Fourth, wall-derived extensin oligomers have been visualized (Cooper *et al.*, 1987). Peroxidase is thought to be the cross-linking agent and peroxidases indeed form extensin dimers *in vitro* (Cooper and Varner, 1984; Everdeen *et al.*, 1988), though in solution the cross-link formed is dityrosine, not isodityrosine (Cooper and Varner, 1984). Peroxidase may exhibit the isodityrosine-synthesizing activity only under suitable conditions *in muro*.

The long-standing hypothesis that extensin network formation is a major determinant of cell expansion (Lamport, 1965) is far from proven. Although in pea epicotyls extensin accumulation coincided with the cessation of growth (Sadava and Chrispeels, 1973), other localization studies argue against a role in expansion. In soybean, high levels of extensin mRNAs and soluble protein were detected in meristematic cells of seedling root and hypocotyl apices and cambium cells of older stems and petioles. Insolubilization of extensins, however, occurred clearly before elongation stopped (Ye and Varner, 1991). The high level of extensin deposited in soybean seed coats (Cassab and Varner, 1987) or secreted in response to wounding and infection (Esquerre-Tugaye and Mazau, 1974) point to a strengthening and protective function unrelated to extensibility as such. Furthermore, since hypocotyl elongation appears to be regulated at the level of the epidermis (Brummell and Hall, 1980; Burstrom, 1979; Rayle *et al.*, 1991), the observed secretion of extensin throughout the shoot apex (Ye and Varner, 1991) would be superfluous. At present, the function of extensin in plant cell wall architecture remains an open question.

Pectin Network

Different kinds of cross-links may give rise to a strong pectin gel. Well-established is the formation of calcium ion bridges between stretches of un-esterified homogalacturonan (Jarvis, 1984). The strength of this type of pectin gel can be regulated by the level of calcium, the pH, or the action of pectin methyl esterase. This apoplastic enzyme is thought to remove methyl groups from esterified pectin blocks delimited by rhamnose residues. Immunolocalization has shown that unesterified homogalacturonan occurs in the middle lamella of the root cortex, with high concentrations around cell junctions at intercellular spaces (Knox *et al.*, 1990). This suggests that calcium pectate gels constitute the glue between cells. Unesterified polygalacturonate also is present close to the plasma membrane (Knox *et al.*, 1990). Since this is thought to be the load-bearing region of the primary cell wall (Taiz, 1984), it is possible that calcium-pectin gels also influence wall extensibility. This is supported by the wall-loosening effect of calcium chelation on soybean hypocotyls (Virk and Cleland, 1988). Other proposed pectin cross-links include ester or secondary amide linkages to other wall polysaccharides and peroxidase-catalysed coupling of phenolic goups (Fry, 1986). The stiffening effect of cross-link formation may be counteracted by esterases or endopolygalacturonase.

The existence of an independent cross-linked pectin network that restricts cell expansion is supported by the behaviour of tomato cells adapted to growth on an inhibitor of cellulose synthesis (Shedletzky *et al.*, 1990). Despite the aberrant walls of these cells - they lack a cellulose/xyloglucan network and appear to consist primarily of pectin with phenolic crosslinks - cell division and controlled extension is possible.

A distinct role of pectins in the cell wall may be in the determination of sieving properties. Permeation and electron microscopy studies both indicate that the size of wall pores increases when pectins are removed from the wall (Baron-Epel *et al.*, 1988; McCann *et al.*, 1990).

Additional Wall Polymers

AGPs. It has been demonstrated with monoclonal antibodies that a family of plasma membrane AGPs is localized at the outer face of the protoplast and the inner 20-30 nm of the cell wall (Knox *et al.*, 1989; Penell *et al.*, 1989). This observation led to the speculation that AGPs may anchor the primary cell wall to the plasma membrane by binding to wall polysaccharides with their lectin domain. There appear to be subtle differences in AGP structure between tissues or regions of developing organs, as indicated by the differential expression AGP epitopes recognized by monoclonal antibodies (Knox *et al.*, 1989, 1990; Penell and Roberts 1990, 1991; Stacey *et al.*, 1990). The bearing of these patterns on morphogenesis and differentiation is unknown. AGPs appear to be turned over rapidly by extracellular hydrolytic enzymes (Fincher *et al.*, 1983; Gibeaut and Carpita, 1991).

GRPs. Though GRPs are deposited in primary cell walls, their function may be related to secondary wall formation in specific cell types. GRP expression has been localized in lignified cells and cells that are going to be lignified (Keller *et al.*, 1988, 1989; Ye and Varner, 1991; Ye *et al.*, 1991). This has led to the view that GRPs, by means of their tyrosine side chains, constitute an initiation site for the deposition of lignin.

Plant cells may have an alternative to limiting their expansion. Proembryogenic masses and the protoderm of developing somatic and zygotic embryos have been found to express the EP2 gene, encoding a lipid transfer protein that is postulated to transport cutin (Martin and Juniper, 1970) monomers to sites of cutin polymerisation (Sterk *et al.*, 1991). The function of a water-repellent cuticle on the outer cell wall of proembryogenic masses and embryos could be to impede water influx, and thereby prevent turgor-driven cell expansion.

Whether the same mechanism functions during early zygotic embryogenesis is at present an open question. In bean, the osmolarity of the liquid endosperm was found to be very high during early embryogenesis but decreased at later stages (Smith, 1973). This would imply that cell wall-controlled limitation of cell expansion in the zygotic embryo is required only at later stages of development. This interpretation is speculative however, because the osmolarities of the developing zygotic carrot embryo and endosperm are not known. Nor is it known whether the peroxidase that prevents cell elongation actually is present in the developing seed.

A mechanism that limits cell expansion may be required also at later stages of somatic embryo development, as indicated by the complementation of arrested globular embryos of ts11 with extracellular proteins. Morphologically, complementation appears to prevent the formation of an aberrant, irregular protodermal layer (De Jong *et al.*, 1992). The protein able to rescue ts11 embryos also has been purified and found to exhibit endochitinase activity (De Jong *et al.*, 1992). Although chitinases are believed generally to act only on cell walls of fungal origin, there may be an as yet unidentified substrate for these enzymes in the plant cell wall.

Tight Cell Adhesion

The second condition for embryogenic development is close contact between daughter cells (Fry, 1990). The coherency of proembryogenic masses and somatic embryos is much higher than that of unorganized clumps of cells and this characteristic would predict altered properties of the polymers that constitute the junctions between cells: the middle lamella and the outer layers of the primary cell wall. Because cell separation can be brought about by pectin degrading enzymes and chelating agents, the middle lamella pectins are generally believed to stick plant cells together. Hence enzymes that enhance pectin gelling, such as pectin methyl esterase or enzymes that may crosslink pectic polysaccharides covalently, are candidates for a promotive role in embryogenesis. However, there is at present no evidence to support this suggestion.

Defense Response

It is of interest to take into consideration that the two secreted proteins identified so far as having promotional effects in carrot somatic embryogenesis have enzyme activities that are also induced when plants are challenged by pathogens or injury (Bowles, 1990; Chibbar *et al.*, 1984; Kragh *et al.*, 1991). The occurrence of these "defense-related" proteins in plant cell suspension culture media generally is interpreted as a wounding response, since tissue coherency is clearly disrupted in cell suspensions. Yet, there may be alternative explanations for the presence of these proteins in culture media, because some of the genes encoding defense-related proteins have been found to be developmentally regulated at the plant level (Lotan *et al.*, 1989; Neale *et al.*, 1990; Ye and Verner, 1991). The effects of the peroxidase and the endochitinase in somatic embryogenesis also support the contention that specific members of the defense-related proteins may function in developmental processes.

Control of Extracellular Protein Function

With the exception of gene products that have to form gradients in the plant body, tissue-specific gene expression only makes sense if the encoded proteins are confined to the cells that produce them. Cytoplasmic protein may be kept easily within the cell boundaries, though some of it will 'escape' through plasmodesmata. For secreted proteins, and more specifically enzymes, the situation is more difficult as they diffuse much more easily through the apoplast to adjacent cells. Hence, additional mechanisms must exist to control the activity of apoplastic enzymes.

In epidermal cells, the asymmetric localization of the nucleus and the cytoplasm is likely to be important to ensure that epidermal wall thickening and cuticle formation occurs only in the cell walls at the plant surface, which is evidently vital to the plant. Still, once secreted the enzymes have to be limited in their diffusion. An obvious way to achieve this is to bind the proteins to the extracellular matrix. The wall-bound forms of EP1 (Van Engelen *et al.*, 1991) may be an example of this type of regulation. Unless the enzyme binds directly to its substrate, this binding is not useful if the enzyme's substrate is immobile because the enzyme has to be able to reach its substrate. More likely than binding in this case, would be degradation of apoplastic enzymes by proteases secreted by the same cell or by neighboring cells. To prevent complete degradation by wall proteases a fairly high continuous synthesis rate of wall enzymes would be required.

If proteolytic breakdown is a general mechanism to control the activity of apoplastic proteins, it probably is disturbed greaten in cell suspension cultures. On one hand secreted proteases, if produced at all under the applied culture conditions, will be highly diluted by the culture medium allowing accumulation of the target proteins. On the other hand, due to the mixing of different cell types, extracellular enzymes may be exposed to proteases they would never encounter *in planta*. This may explain why the 31/32 kD EP1 proteins are found only in tissue culture and also why EP1 proteins accumulate to normal levels in a suspension culture derived from leaf tissue which, by itself appears to degrade its EP1 proteins (Van Engelen *et al.*, 1991).

Somatic Embryogenesis: Development in a Hostile Environment?

In the preceding paragraphs the central role of the plant cell wall in the generation of plant form has been emphasized. The view has been put forward that secreted enzymes act as regulators of morphogenesis by changing the chemical structure of cell wall polymers and consequently, the mechanical properties of the wall as a whole. When this view is applied to somatic embryogenesis in cell suspension cultures, there is enough reason to be surprised that embryos develop at all in such a system. In most embryogenic cell suspensions, embryogenic cells are outnumbered by far by non-embryogenic cells which are characterized by extensive expansion and loose attachment to other cells. If these properties of non-embryogenic cells are mediated by secreted enzymes, these enzymes will accumulate in the culture medium especially in high density cultures. Thus, we might expect from this that conditioned culture medium is essentially hostile toward the development of proembryogenic masses as they are immersed in a solution of enzymes that promote cell expansion and/or loosen cell contacts. However, experience tells us that suspension culture conditions do permit the development of embryogenic cells, i.e. under identical growth conditions some cells expand while others do not. This phenomenon may be explained in two ways.

First, the enzymes may simply be diluted too much in the culture medium to reach a critical level. The highest concentration of an expansion-promoting enzyme clearly would be found in the walls of the non-embryogenic cells themselves where they must be active. The concentration probably decreases sharply towards the medium and the effect of such an enzyme on the cell wall of an embryogenic cell then might be too minor to alter its propertiessignificantly. In this respect, it is worthwhile to note that in carrot, somatic embryo formation from proembryogenic masses involves enrichment for embryogenic cells followed by an at least 20 fold reduction in cell density. In addition, the presence of proteins inhibitory to somatic embryogenesis may also be reduced under embryo-forming conditions. One example of such proteins are those identified by Gavish *et al.* (1992) in *Citrus*.

Second, the structure of the wall of an embryogenic cell may *a priori* be such that the enzymes have no access to their substrates. In addition to a restriction in water-uptake, this might be a second function of the presence of a hydrophobic layer on the surface of proembryogenic masses and somatic embryos (Sterk *et al.*, 1991). The low amounts of protein secreted by proembryogenic masses and somatic embryos when compared to non-embryogenic suspension cells (unpublished observations) indicates that this barrier may also function in the opposite direction, limiting the diffusion of the proteins secreted by these structures.

Finally, because at least two secreted proteins were shown to actually promote somatic embryogenesis in an arrested situation (Cordewener *et al.*, 1991; De Jong, 1992) the relationships between cultured plant cells and their surrounding medium apparently involves a range of putative cell wall enzymes. To understand the interactions between secreted (cell wall) glycoproteins, their putative substrates in the cell walls and the implications of these interactions for plant development in precise biochemical detail will no doubt turn out to be an extremely complex undertaking.

ACKNOWLEDGEMENTS

Original research of our group is supported by the Netherlands Organization of Scientific Research and the European Community Program BRIDGE.

REFERENCES

Ammirato PV (1983) Embryogenesis, *In*: Handbook of Plant Cell Culture, vol 1, Evans DA, Sharp WR, Yamada Y (eds), Macmillan, New York, pp 88-123.

Angelini R, Federico R (1989) Histochemical evidence of polyamine oxidation and generation of hydrogen peroxide in the cell wall. *J Plant Physiol* 135:212-217.

Averyhart-Fullard V, Datta K, Marcus A (1988) A hydroxyproline-rich protein in the soybean cell wall. *Proc Natl Acad Sci USA* 85:1082-1085.

Backs-Husemann D, Reinert J (1970) Embryobildung durch isolierte Einzelzellen aus Gewebekulturen von *Daucus carota*. *Protoplasma* 70:49-60.

Barondes SH (1988) Bifunctional properties of lectins: lectins redefined. *Trends Biochem Sci* 13:480-482.

Baron-Epel O, Gharyal PK, Schindler M (1988) Pectins as mediators of wall porosity in soybean cells. *Planta* 175:389-395.

Bellincampi D, Morpurgo G (1987) Conditioning factor affecting growth in plant cells in culture. *Plant Sci* 51:83-91.

Biggs KJ, Fry SC (1990) Solubilization of covalently bound extensin from *Capsicum* cell walls. *Plant Physiol* 92:197-204.

Bohlmann H, Clausen S, Behnke S, Giese H, Hiller C, Reimann-Phillipp U, Schrader G, Barkholt V, Apel K (1988) Leaf-specific thionins of barley - a novel class of cell wall proteins toxic to plant-pathogenic fungi and possibly involved in the defence mechanisms of plants. *EMBO J* 7:1559-1565.

Bowles DJ (1990) Defense-related proteins in higher plants. *Annu Rev Biochem* 59:873-907.

Branca C, De Lorenzo G, Cervone F (1988) Competitive inhibition of the auxin-induced elongation by a-D-oligogalacturonides in pea stem segments. *Physiol Plant* 72:499-504.

Brummell DA, Hall JL (1980) The role of the epidermis in auxin-induced and fusicoccin-induced elongation of *Pisum sativum* stem segments. *Planta* 150:371-379.

BurstromHG (1979) In search of a plant growth paradigm. *Am J Bot* 66: 98-104.

Cassab GI, Varner JE (1988) Cell wall proteins. *Annu Rev Plant Physiol Plant Mol Biol* 39:321-353.

Cassab GI, Varner JE (1987) Immunocytolocalization of extensin in developing soybean seed coats by immunogold-silver staining and by tissue printing on nitrocellulose paper. *J Cell Biol* 105:2581-2588.

Catesson A-M, Imberty A, Goldberg R, Czaninski Y (1986) Nature, localization and specificity of peroxidases involved in lignification processes *In:* Molecular and Physiological Aspects of Plant Peroxidases, Greppin H, Penel C, Gaspar T (eds), University of Geneva, pp 189-198.

Chen J, Varner JE (1985) Isolation and characterization of cDNA clones for carrot extensin and a proline-rich 33 kDa protein. *Proc Natl Acad Sci USA* 82:4399-4403.

Chibbar RN, Cella R, Albani D, Van Huystee RB (1984) The growth and peroxidase synthesis by two carrot cell lines. *J Exp Bot* 35:1846-1852.

Condit CH, Meagher RB (1986) A gene encoding a novel glycine-rich structural protein of petunia. *Nature* 323:178-181.

Cooper JB, Varner JE (1984) Cross-linking of soluble extensin in isolated cell walls. *Plant Physiol* 76:414-417.

Cooper JB, Chen JA, Van Holst GJ, Varner JE (1987) Hydroxyproline-rich glycoproteins of plant cell walls. *Trends Biochem Sci* 12: 24-27.

Cordewener J, Booij H, Van der Zandt H, Van Engelen F, Van Kammen A, De Vries S (1991) Tunicamycin-inhibited carrot somatic embryogenesis can be restored by secreted cationic peroxidase isoenzymes. *Planta* 184:478-486.

Darvill A, McNeil M, Albersheim P, Delmer DP (1980) The primary cell walls of flowering plants. *In:* The Biochemistry of Plants, vol 1, Tolbert NE (ed), Academic press, New York, pp 91-162.

Darvill AG, Albersheim P, McNeil M, Lau JM, York WS, Stevenson TT, Thomas J, Doares S, Gollin DJ, Chelf P, Davis K (1985) Structure and function of plant cell wall polysaccharides. *J Cell Sci Suppl* 2:203-217.

Kragh KM, Jacobsen S, Mikkelsen JD, Nielsen KA (1991) Purification and characterization of three chitinases and one β-1,3-glucanase accumulating in the medium of cell suspension cultures of barley (*Hordeum vulgare* L). *Plant Sci* 76:65-78.

Lagrimini LM, Bradford S, Rothstein S (1990) Peroxidase-induced wilting in transgenic tobacco plants. *Plant Cell* 2:7-18.

Lamport DTA (1965) The protein component of primary cell walls. *In*: Advances in Botanical Research, Preston RD (ed), vol 2, Academic press, New York, pp 151-218.

Li X, Kieliszewski M, Lamport DTA (1990) A chenopod extensin lacks repetitive tetrahydroxyproline blocks. *Plant Physiol* 92:327-333.

LoSchiavo F, Giuliano G, de Vries SC, Genga A, Bollini R, Pitto L, Cozzani F, Nuti-ronchi V, Terzi M (1990) A carrot cell variant temperature-sensitive for somatic embryogenesis reveals a defect in the glycosylation of extracellular proteins. *Mol Gen Genet* 223:385-393.

LoSchiavo F Quesada-Allue LA, Sung ZR (1986) Tunicamycin affects somatic embryogenesis but not cell proliferation of carrot. *Plant Sci* 44:65-71.

Lotan T, Ori N, Fluhr R (1989) Pathogenesis-related proteins are developmentally regulated in tobacco flowers. *Plant Cell* 1:881-887.

MacKay AL, Wallace JC, Sasaki K, Taylor IEP (1988) Investigation of the physical structure of the primary plant cell wall by proton magnetic resonance. *Biochemistry* 27: 1467-1473.

Martin JT, Juniper BE (1970) The Cuticles of Plants. Edward Arnold, London.

McCann MC, Roberts K (1991) Architecture of the primary cell wall. *In*: The Cytoskeletal Basis of Plant Growth and Form. Lloyd CW (ed), Academic Press, London, pp 109-128.

McCann MC, Wells B, Roberts K (1990) Direct visualization of cross-links in the primary cell wall. *J Cell Sci* 96:323-334.

McDougall GJ, Fry SC (1988) Inhibition of auxin-stimulated growth of pea stem segments by a specific nonasaccharide of xyloglucan. *Planta* 175:412-416.

McDougall GJ, Fry SC (1990) Xyloglucan oligosaccharides promote growth and activate cellulase: evidence for a role of cellulase in cell expansion. *Plant Physiol* 93:1042-1048.

McNeil M, Darvill AG, Fry SC, Albersheim P (1984) Structure and function of the primary cell walls of plants. *Annu Rev Biochem* 53:625-663.

Mort AJ, Lamport DTA (1977) Anhydrous hydrogen fluoride deglycosylates glycoproteins. *Anal Biochem* 82: 289-309.

Neale AD, Wahleithner JA, Lund M, Bonnett HT, Kelly A, Meeks-Wagner DR, Peacock WJ, Dennis ES (1990) Chitinase, β-1,3-glucanase, osmotin and extensin are expressed in tobacco explants during flower formation. *Plant Cell* 2:673-684.

Nomura K, Komamine A (1985) Identification and isolation of single cells that produce somatic embryos at a high frequency in a carrot suspension culture. *Plant Physiol* 79:988-991.

Northcote DH, Davey R, Lay J (1989) Use of antisera to localize callose, xylan and arabinogalactan in the cell-plate, primary and secondary walls of plant cells. *Planta* 178:353-366.

Nothnagel E, McNeil M, Albersheim P, Dell A (1983) Host-pathogen interactions XXII A galacturonic acid oligosaccharide from plant cell walls elicits phytoalexins.

Osteryoung KW, Toenjes K, Hall B, Winkler V, Bennett AB (1990) Analysis of tomato polygalacturonase expression in transgenic tobacco. *Plant Cell* 2:1239-1248.

Pennell RI, Knox JP, Scofield GN, Selvendran RR, Roberts K (1989) A family of abundant plasma membrane-associated glycoproteins related to the arabinogalactan proteins is unique to flowering plants. *J Cell Biol* 108:1967-1977.

Pennell RI, Roberts K (1990) Sexual development in the pea is presaged by altered expression of arabinogalactan protein. *Nature* 344:547-549.

Pennell RI, Janniche L, Kjellbom P, Scofield GN, Peart JM, Roberts K (1991) Developmental regulation of a plasma membrane arabinogalactan epitope in oilseed rape flowers. *Plant Cell* 3:1317-1326.

Pressey R (1991) Oxidized oligogalacturonides activate the oxidation of indoleacetic acid by peroxidase. *Plant Physiol* 96:1167-1170.

Rayle DL, Nowbar S, Cleland RE (1991) The epidermis of the pea epicotyl is not a unique target tissue for auxin-induced growth. *Plant Physiol* 97:449-451.

Reinert J (1959) Über die Kontrolle der Morphogenese und die Induktion von Adventivembryonen an Gewebekulturen aus Karotten. *Planta* 53: 318-333.

Roberts K (1989) The plant extracellular matrix. *Curr Opin Cell Biol* 1:1020-1027.

Sack F (1991) What is a plant cell? *Continued Plant Cell* 3: 844.

Sadava D, Chrispeels MJ (1973) Hydroxyproline-rich cell wall protein extensin role in the cessation of elongation in excised pea epicotyls. *Dev Biol* 30:49-55.

Selvendran RR (1985) Developments in the chemistry and biochemistry of pectic and hemicellulosic polymers. *J Cell Sci Suppl* 2:51-88.

Shedletzky E, Shmuel M, Delmer DP, Lamport DTA (1990) Adaptation and growth of tomato cells on the herbicide 2,6-dichlorobenzonitrile leads to production of unique cell walls virtually lacking a cellulose-xyloglucan network. *Plant Physiol* 94:980-987.

Smith JG (1973) Embryo development in *Phaseolus vulgaris*. II. Analysis of selected inorganic ions, ammonia, organic acids, amino acids, and sugars in the endosperm liquid. *Plant Physiol* 51:454-458.

Smith RC, Fry SC (1991) Endotransglycosylation of xyloglucans in plant cell supension cultures. *Biochem J* 279:529-535.

Spangenberg G, Koop H-U, Schweiger H-G (1985) Different types of protoplasts from *Brassica napus* L: analysis of conditioning effects at the single cell level. *Eur J Cell Biol* 39:41-45.

Stacey NJ, Roberts K, Knox JP (1990) Patterns of expression of the JIM4 arabinogalactan-protein epitope in cell cultures and during somatic embryogenesis in *Daucus carota* L. *Planta* 180:285-292.

Staehelin A (1991) What is a plant cell? A response. *Plant Cell* 3:553.

Stafford H (1991) What is a plant cell? *Plant Cell* 3:331.

Steeves TA, Sussex I (1989) Patterns in Plant Development, Cambridge University Press.

Sterk P, Booij H, Schellekens GA, Van Kammen A, De Vries SC (1991) Cell-specific expression of the carrot EP2 lipid transfer protein gene. *Plant Cell* 3:907-921.

Stevenson TT, McNeil M, Darvill AG, Albersheim P (1986) Structure of plant cell walls. XVIII. An analysis of the extracellular polysaccharides of suspension cultured sycamore cells. *Plant Physiol* 80: 1012-1019.

Steward FC, Mapes MO, Smith J (1958) Growth and organized development of cultured cells. I. Growth and division of freely suspended cells. *Am J Bot* 45:693-703.

Stiefel V, Péres-Grau L, Albericio F, Giralt E, Ruiz-Avila E, Ludevid MD, Puigdomènech P (1988) Molecular cloning of cDNAs encoding a putative cell wall protein from *Zea mays* and immunological identification of related polypeptides. *Plant Mol Biol* 11: 483-493.

Stuart DA, Varner JE (1980) Purification and characterization of a salt-extractable hydroxyproline-rich glycoprotein from aerated carrot discs. *Plant Physiol* 66:787-792.

Taiz L (1984) Plant cell expansion: regulation of cell wall mechanical properties. *Annu Rev Plant Physiol* 35:585-657.

Talbott LD, Ray PM (1992) Molecular size and separability features of pea cell wall polysaccharides Implications for models of primary wall structure. *Plant Physiol* 98:357-368.

Talbott LD, Ray PM (1992) Changes in molecular size of previously deposited and newly synthesized pea cell wall matrix polysaccharides. *Plant Physiol* 98:369-379.

Tierney ML, Varner JE (1987) The extensins. *Plant Physiol* 84:1-2.

Tuerana CE, Taylor AJ, Mitchell JR (1982) Evaluation of a method for determining the free carboxyl group distribution in pectins. *Carbohydr Polymers* 2:193-203.

Van Engelen FA, Sterk P, Booij H, Cordewener JHG, Rook W, Van Kammen A, De Vries SC (1991) Heterogeneity and cell type-specific localization of a cell wall glycoprotein from carrot suspension cells. *Plant Physiol* 96:705-712.

Varner JE, Cassab G (1986) A new protein in petunia. *Nature* 323:110.

Varner JE, Lin L-S (1989) Plant cell wall architecture. *Cell* 56:231-239.

Virk SS, Cleland RE (1988) Calcium and the mechanical properties of soybean hypocotyl cell walls: possible role of calcium and protons in cell-wall loosening. *Planta* 176:60-67.

Xu ZH, Huang B, Sunderland N (1981) Culture of barley anthers in conditioned media. *J Exp Bot* 32:767-778.

Xu ZH, Sunderland N (1981) Glutamine, inositol, and conditioning factor in the production of barley pollen callus *in vitro*. *Plant Sci L* 23:161-168.

Ye Z-H, Song Y-R, Marcus A, Varner JE (1991) Comparative localization of three classes of cell wall proteins. *Plant J* 1:175-183.

Ye Z-H, Varner JE (1991) Tissue-specific expression of cell wall proteins in developing soybean tissues. *Plant Cell* 3:23-37.

AN ASSESSMENT OF POSSIBLE FACTORS CONTRIBUTING TO RECALCITRANCE OF PLANT PROTOPLASTS

Kalliopi A. Roubelakis-Angelakis

Department of Biology, University of Crete
P.O. Box 1470, 71110 Heraklion, GREECE

INTRODUCTION

Plant tissue culture techniques, including protoplast technology, have been based on a unique property of plant cells, totipotency. This property enables plant cells to change the order of the developmental expression of their genetic dictionary and instead of expressing genes normally involved in the developmental processes pertinent to the tissue the cells belonged to, they become competent to express genes regulating mitotic activity.

Hanstein (1880) was probably the first to use the term protoplast (from the Greek proto and plastos, which means first created) for plant cells lacking cell walls and to isolate protoplasts from cells of *Vacuheria*. In 1892, Klercker was the first to isolate protoplasts following plasmolysis of leaf tissue cells of *Stratiotes aloides*. A significant development was achieved by Giaja in 1919, when he isolated protoplasts from yeasts by digesting the cell walls with gastric juice from the snail *Helix pomatia*. Large scale isolation of protoplasts was possible only when hydrolytic enzymes were employed in the 60's (Cocking, 1961).

Plant protoplasts offer an excellent model for physiological, biochemical, molecular and virological studies. Further, protoplasts especially from haploid plants can be handled like microbes and are well suited to mutagenic treatments. Also, protoplast technology can offer solutions for improving plants by either DNA transfer or by somatic hybridization. A prerequisite for the production of transgenic plants by protoplast technology is the availability of a method to regenerate a complete plant from the altered cell. Despite extensive studies over the past thirty years, regeneration from protoplasts has been obtained in a limited number of plant species only. Protoplasts from the most important agricultural species, such as most cereals, legumes, and woody perennials, show recalcitrance to complete plant regeneration.

Morphogenesis in Plants, Edited by K.A. Roubelakis-Angelakis
and K. Tran Thanh Van, Plenum Press, New York, 1993

infectation, slicing, etc.). For instance, in *Pennisetum americanum*, protoplasts from callus derived from immature embryos were regenerated into plants, whereas protoplasts from other sources were unsuccessful (Vasil and Vasil, 1980). Also, grapevine (*Vitis vinifera* L.) mesophyll protoplasts from *in vitro* cultured plants showed higher viability rates during culture and were less recalcitrant to induction of cell division than protoplasts from fully developed leaves from either greenhouse or field grown plants (Theodoropoulos and Roubelakis-Angelakis, 1990).

The conditions of the isolation procedure which mostly affect the overall quality of protoplasts are the kind, the purity and the concentration of the cell wall degrading enzymes, the duration of the enzymic digestion, temperature and the light conditions. Also, the correct kind of plasmolyticum and its concentration, the pH of the incubation medium and other substances (e.g. potassium dextran sulfate, Ca^{2+}, Mg^{2+} ions etc.) added to the medium affect the overall quality of protoplasts and their regeneration potential.

Recently, Siminis *et al.* (1993b) found that tobacco protoplasts exhibited high plating efficiencies when the duration of enzyme action (maceration) was 4h, whereas when this period was extended to 18h, the plating efficiency was reduced dramatically. Also, viral infection of donor plants alter the composition of cell walls (Wang *et al.*, 1986) and the use of actinomycin during culture may cause structural aberrations in the cells (Nagata and Takebe, 1970). These alterations can be eliminated by using virus-free plant material from *in vitro* grown plantlets (Theodoropoulos and Roubelakis-Angelakis, 1990). In general, *in vitro* culture of plant material has made controlled application of environmental factors possible to investigate their significance for the regenerative potency of the isolated protoplasts (Maheshwari *et al.*, 1986). For instance, increased regeneration frequencies were obtained with protoplasts from moss protonema grown at low calcium (Saxena and Rashid, 1981).

Often papers have detailed isolation procedures from plants grown under normal greenhouse conditions or have mentioned that the techniques were usefull only during a portion of the year. Inevitably, such methods fail the test of reproducibility in terms of either protoplast yield, viability or ability to proliferate in culture. Each of these criteria can be independent considerations, but all must be satisfied before efficient colony formation is possible (Shepard, 1981). Meyer *et al.* (1993) state that isolation has a far more important impact than the composition of the culture medium.

CULTURE CONDITIONS

The culture conditions which significantly affect the viability and the potential for regeneration of plant protoplasts are the physical and the chemical environment. The quality and intensity of light, temperature, probably gaseous concentrations in the microenvironment of protoplast culture and the density of protoplasts are crucial parameters of the physical environment. The composition of culture medium, such as the carbon source, vitamins, osmoticum, plant growth regulators, other additives and the pH determine the chemical environment of protoplast culture.

Optimization of culture conditions of protoplasts requires long and tedious experimental work due to the complexity of the composition of culture medium. The value of pH is very crucial to the survival of protoplasts and it must be noticed that during culture, pH decreases substantially. Also, osmolality is another significant parameter. For the recalcitrant grapevine protoplasts, optimum osmolality for higher viability rates is 7.30 to 9.30 mosmol Kg^{-1} (Katsirdakis and Roubelakis-Angelakis, 1992a). In the same system 6-benzylaminopurine and naphthalene acetic acid at $2.3 \cdot 10^{-6}$ and $10\text{-}15 \cdot 10^{-6}$ M, respectively result in higher viability rates and elongation of protoplasts (Katsirdakis and Roubelakis-Angelakis, 1992a).

Lack of success in plant regeneration from protoplasts does not necessarily imply that the material is recalcitrant. Careful search for the optimum culture conditions for each plant species, and even for each cultivar may result in higher viability rates and occasionally in success. Thus, whether or not recalcitrance is genetically dictated, proper manipulation of the culture conditions and especially of the chemical environment may at least partially overcome the problem.

GENE EXPRESSION - DNA METHYLATION

Gene expression is certainly among the most significant factors which seems to contribute to recalcitrance of plant protoplasts. In this volume a detailed review concerning modified gene expressions in freshly isolated and cultured plant protoplasts is presented (Meyer *et al.*, 1993). Changes in protein synthesis takes place early during or after protoplast isolation. The mRNAs encoded by the nuclear genes decline during isolation (Lett *et al.*, 1980; Vernet *et al.*, 1982), while those corresponding to chloroplast genes continue to be expressed at a high level for at least two days (Criqui *et al.*, 1992). Rapid repression of the nuclear encoded photosynthetic genes in mesophyll protoplasts is primarily induced by the mechanical wounding stress prior to maceration.

The *de novo* synthesis of proteins in cultured dividing (*Nicotiana tabacum*) and non-dividing (*Vitis vinifera*) protoplasts exhibited three and two peaks, respectively, coinciding with isolation (stress proteins), cell wall synthesis and induction of mitosis. In recalcitrant protoplasts the third peak was lacking (Katsirdakis and Roubelakis-Angelakis, 1992a). Also, incorporation of tritiated uridine and thymidine in cultured protoplasts was higher in tobacco than in grapevine (Fig. 3).

Recently, results from a comparative study of the profile of the *de novo* synthesized polypeptides in same protoplasts has shown that the number of polypeptides which changed in tobacco protoplasts was almost double compared to grapevine protoplasts after 96h in culture. The bulk of the proteins which increased in grapevine protoplasts was of relatively low molecular weight, whereas in tobacco, the molecular weight varied. Also the polypeptides which increased in grapevine protoplasts were only in the acidic region, when fractionated on IEF gels, whereas in tobacco protoplasts they were both acidic and basic. The basic proteins of cultured tobacco protoplasts resembled the PR proteins detected by Grosset *et al.* (1990) in both their pIs and molecular weights. Finally, the labeled polypeptides which declined after 96h in culture were mostly in the basic region for the grapevine and in the acidic region for tobacco (Siminis *et al.*, 1993a).

Loss of competence by recalcitrant cells is linked to their genetic/epigenetic status

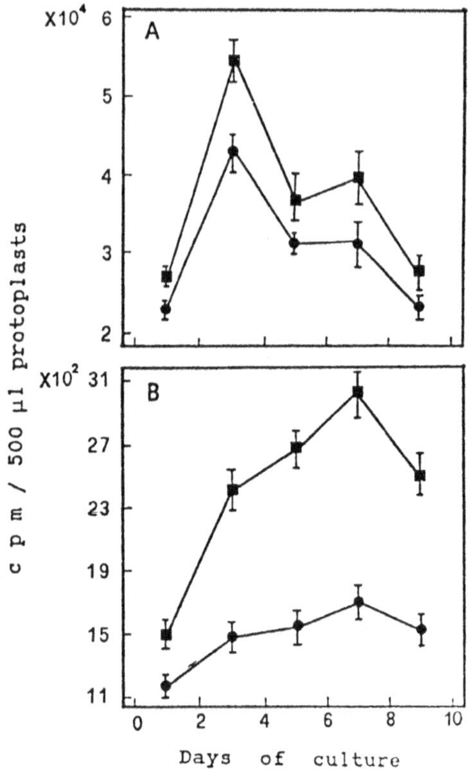

Figure 3. Incorporation of labeled uridine (upper) and thymidine (lower) in *Nicotiana tabacum* (■) and *Vitis vinifera* (●) protoplasts (Katsirdakis and Roubelakis-Angelakis, unpublished).

(Mc Clelland and Nelson, 1985; Meins, 1983). It has been suggested that recalcitrance is linked to changes in gene expression, which may be controlled by quantitative or qualitative changes in DNA. One qualitative change in DNA which is believed to participate in the regulation of transcription and differentiation is the extent of DNA methylation.

Results from studies on mammallian cells support that methylation may participate in gene regulation. Changes in the methylation pattern of genes during differentiation are correlated with changes in gene expression, and methylated genes introduced into cells are not expressed when compared with unmethylated genes (Adams and Burdon, 1985; Herburn *et al.*, 1987; Heseman and Schroder, 1982).

Comparative data on DNA methylation in recalcitrant and regenerating protoplasts are lacking. Data from the grass *Pennisetum purpureum* indicate that DNA is highly methylated and that loss of embryogenic competence is not linked to a gross alteration in DNA methylation (Morrish and Vasil, 1989). The authors suggest that cell competence may be controlled by changes in DNA methylation patterns, which were not detected by the techniques employed; in other plants putative correlation between methylation and gene expression has been found by using specific gene probes (Nick *et al.*, 1986; Roberts, 1980;

Schwartz and Dennis, 1986). Thus, the elucidation of the putative role played by DNA methylation in protoplast recalcitrance requires the identification and cloning of specific genes which could be used as specific probes.

PLASMA MEMBRANE FUNCTIONING

The plasma membrane is the major barrier separating the cells from their environment and its importance in the regulation of a number of processes at the cellular level has long been recognized. During the enzymatic digestion of cell walls, the plasma membrane may be seriously injured; this injury may in turn affect the structure and functioning of the plasma membrane during the subsequent culture of protoplasts. Furthermore, there is increasing evidence that free radical-induced lipid peroxidation leads to accumulation of peroxidized lipids in cell membranes and loss of specialized membrane functions.

In an effort to test the integrity and the functioning of plasma membrane, the uptake characteristics of sugars and amino acids by mesophyll protoplasts from a recalcitrant plant species, the grapevine (*Vitis vinifera* L.), were studied. Uptake of ^{14}C-L-arginine was linear up to at least 60 min. The uptake rate from a 2 μM external concentration was 100 pmol per 10^6 viable protoplasts per h, which corresponds to an intracellular concentration of labeled compound of 8.3 μM. The uptake rate was independent of light conditions during the assay, whereas preincubation of protoplasts in the dark for 24h resulted in a decrease of uptake rate by 70% (Roubelakis-Angelakis and Theodoropoulos, 1990). The presence of Ca^{2+} did not affect uptake rates. Kinetics analysis revealed a biphasic uptake curve. The high affinity component had a Km of 2.2 mM. Two carrier systems, one for basic and neutral and one for acidic amino acids were identified. Use of inhibitors revealed that those associated with ATP metabolism inhibited arginine uptake; more specifically, the proton motive force appeared to be the predominant energy source. Metabolic products of labeled arginine were consistent with the operation of the Krebs-Henseleit cycle (Roubelakis and Kliewer, 1978a,b,c; Theodoropoulos and Roubelakis- Angelakis, 1989).

Cultured protoplasts rapidly deplete glucose from the culture medium and success in regenerating plants from protoplasts requires a continuing uptake capacity of substances by the cultures. Recalcitrant *Vitis vinifera* L. protoplasts showed a rate of glucose uptake from a 2.0 μM external solution of 130 pmol per 10 viable protoplasts per h and was linear for 60 min. Glucose uptake in the dark was lower compared to that in light and azide significantly inhibited the process (Roubelakis-Angelakis and Theodoropoulos 1990; Theodoropoulos and Roubelakis-Angelakis, 1990). The transport system was hexose specific and the stereospecifity was closely related to carbon-1 of the glucose molecule. Glucose structural analogues were not metabolized beyond the stage of phosphorylation and differences between these compounds and glucose were observed in their transport rates and in their carrier specificity. Uptake of labeled glucose by grapevine protoplasts was linear for concentrations higher than 1.5 μM; at lower concentrations a saturating tendency was observed. The carrier was driven by the proton motive force and the substrate entered the cell probably in an unchanged form. Efflux studies were not useful as an indication of the rate of metabolism or assimilation of transported compounds in recalcitrant protoplasts (Theodoropoulos and

was more pronounced in tobacco protoplasts. Also, in both systems α-tubulin was localized in the same locations as actin (Katsirdakis, 1991).

Calmodulin (CaM) is a calcium-binding protein that has been implicated in the regulation of the assembly and disassembly of cytoskeletal structures. In animal systems, high levels of cellular CaM have been reported in cells that fall into two general categories: proliferating cells in regions of high mitotic activity and cells involved in specialized rapid signalling (Klee and Vanaman, 1982). Recently, CaM levels, on a per protein basis, were found to be higher in guard cell and epidermal cell protoplasts than mesophyll cell protoplasts (Ling and Assmann, 1992). Also, during the exponential phase of carrot suspension culture growth when mitotic activity and oxidative respiration were maximal, CaM mRNA levels were 4- to 5-fold higher than they were during the later stages of culture growth, when respiration rates were lower and growth was primarily by cell expansion. Net CaM polypeptide synthesis paralleled the changes in CaM steady-state mRNA level. As a consequence, net CaM polypeptide synthesis declined 5- to 10- fold during the later stages of culture growth. CaM polypeptide levels, in contrast to its net synthesis, remained relatively constant during the exponential phases of the culture growth cycle and increased during the later stages of culture growth. These data suggest increased CaM polypeptide turnover associated with periods of rapid cell proliferation and high levels of respiration (Perera and Zielinski, 1992).

The questions on the possible role of actin microfilament network and microtubules in cell wall structure and of CaM, calcium concentrations and cell polarity during protoplast isolation and regeneration remain open.

OXIDATIVE STRESS

There is no doubt that during the isolation of plant protoplasts, serious stress conditions and wound-induced reactions take place. Among other side effects, there is a possibility that the homeostatic mechanisms of oxygen metabolism are not maintained and/or the free radical scavenging mechanisms do not efficiently protect protoplasts from oxidative stress.

The superoxide radical ($O_2^{\cdot-}$), the hydroxyl radical (·OH), hydrogen peroxide and the activated singlet oxygen ($^1O_2$) are the most important damaging reactants (Fridovich, 1986). Their increased production during stress and aging leads to the formation of lipid radicals and peroxides. The latter are potentially very toxic to the cell as their aldehyde breakdown products may crosslink with important cellular macromolecules, such as DNA and proteins. Many of the secondary lipid peroxidation products are fluorescent; in mammallian cells the fluorescent compounds formed by the cross linking of amino groups with aldehydes are collectively called lipofuchsin pigments (Donats, 1981). These compounds have been often used as indicators of oxidative stress; they are called Schiff's bases (Benson, 1990).

Antioxidants, such as glutathione, ascorbate and α-tocopherol, are the endogenous protectants of oxygen toxicity. Also, at least three enzymic mechanisms operate in the cells against oxidative damages: peroxidase (EC 1.11.1.7.) and catalases (EC 1.11.1.6.) remove hydrogen peroxide, whereas superoxide dismutase (SOD, EC 1.15.1.1.) converts superoxide to hydrogen peroxide.

A comparative study of peroxidases, catalase and SOD in protoplasts of the recalcitrant *Vitis vinifera* L. and the regenerating *Nicotiana tabacum* in our laboratory has shown that peroxidase activity was low in freshly isolated protoplasts in both species (0.6 and 1.6 units per mg protein, respectively); during the subsequent 8 days of culture peroxidase activity in protoplasts increased 8- and 120-fold, in *Vitis* and *Nicotiana*, respectively (Fig. 4). There were also significant differences in the isoperoxidase profile between the two species. In *Nicotiana*

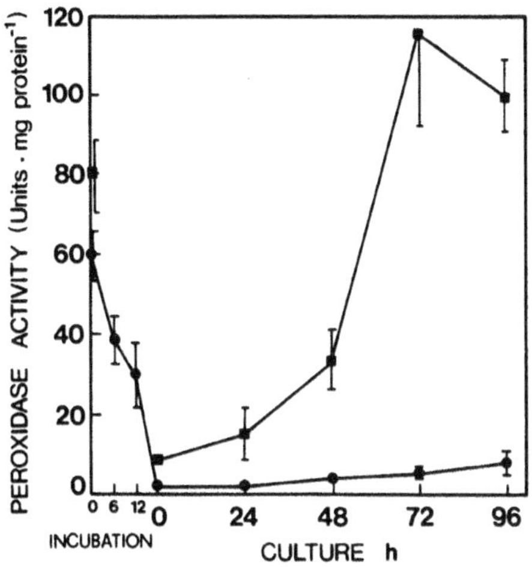

Figure 4. Specific activity of peroxidase in cultured *Nicotiana tabacum* (■) and *Vitis vinifera* (●) protoplasts (Siminis, Kanellis, and Roubelakis-Angelakis, unpublished).

protoplasts, the expression of basic isoperoxidases was correlated with cell division, whereas in recalcitrant grapevine protoplasts their absence was correlated with their inability to enter the cell cycle (Siminis *et al.*, 1993b). Also, the lipid peroxidation products in *Vitis* protoplasts during culture were significantly higher than in *Nicotiana* protoplasts (Fig. 5).

Catalase activity, isoforms and immunoreactive protein were studied during protoplast isolation and culture. At zero time (isolation) activity of catalase in tobacco protoplasts was almost 2-fold greater than in *Vitis* protoplasts. Thereafter, enzyme activity increased significantly during the first 24h of culture in *Vitis* and remained at high levels during the subsequent days of culture; in tobacco protoplasts with high plating efficiency enzyme activity showed a continuing decline during the same period (Fig. 6). When the duration of isolation increased to 18h, the plating efficiency was less than 5%, and catalase activity rapidly decreased after a short increase during the isolation process. The high activity in *Vitis* was due to a newly synthesized 57 kD subunit of catalase giving rise to the more acidic isoforms, without altering their number. In contrast, in *Nicotiana* a shift in catalase expression was observed; the 57 kD subunit, mainly expressed till this time, disappeared and only the 56 kD subunit remained in increasing amounts, giving rise to the more acidic isoforms while the basic ones where eliminated. When the regenerating potential of tobacco protoplasts was

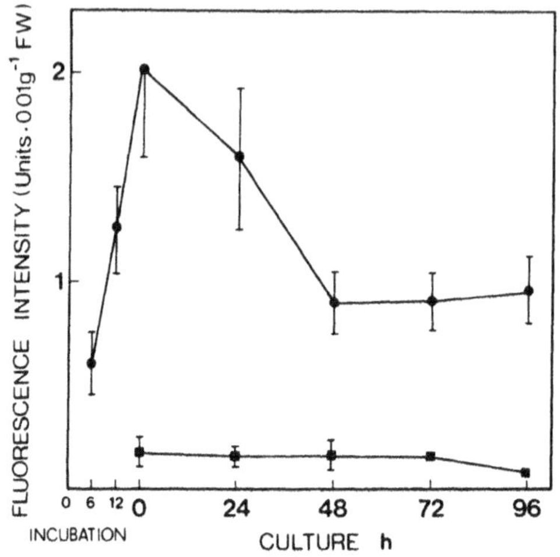

Figure 5. Lipid peroxidation products in cultured *Nicotiana tabacum* (■) and *Vitis vinifera* (●) protoplasts (Siminis, Kanellis, and Roubelakis-Angelakis, unpublished).

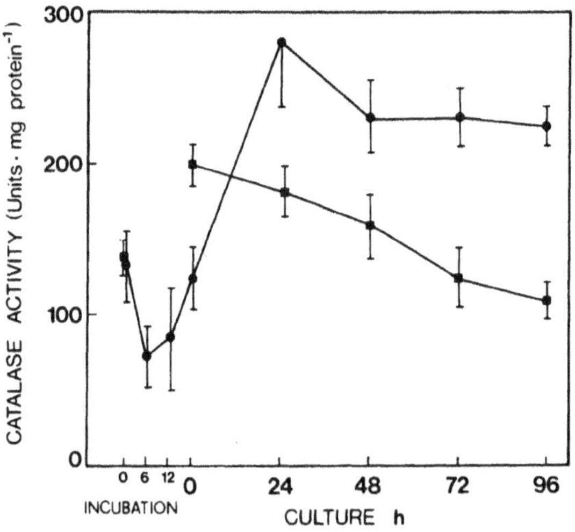

Figure 6. Specific activity of catalase in cultured *Nicotiana tabacum* (■) and *Vitis vinifera* (●) protoplasts (Siminis, Kanellis, and Roubelakis-Angelakis, unpublished).

very low, the 56 kD subunit disappeared after 96h in culture, and only the 57 kD subunit remained in low amounts in the basic isoforms present in the original protoplasts (Siminis *et al.*, 1990, 1992, 1993b). Significant differences were found in the catalase isoforms by immunoblotting after IEF separation of protoplast proteins. In dividing *Nicotiana* protoplasts during culture, the acidic catalase isoenzymes increased, whereas the basic ones decreased. The opposite was true for non-dividing grapevine protoplasts, where the acidic isocatalases consisted mostly of 57 kD, while the basic ones consisted mostly of 56 kD subunit.

Three classes of SOD have been identified based on the metal present at the active site: copper/zinc (Cu/Zn SOD), iron (FeSOD) and manganese (MnSOD) (Fridovich, 1986). The iron enzyme is present in procaryotes and within the plastids of some plants. The MnSOD is widely distributed among procaryotic and eucaryotic organisms and in eucaryotes, it is mostly found in the mitochondrial matrix. These two enzymes are likely to have arisen from the same ancestral enzyme, whereas the Cu/ZnSOD has no sequence similarity to them and appears to have evolved separately in eucaryotes. It is mostly cytosolic and in some plants a chloroplastic isoform is present (for review see Bannister *et al.*, 1987). Recently, it was proposed that these SODs can be used as molecular probes to study the oxidative stress that occurs specifically in plant cytoplasm, mitochondria, and chloroplasts as a result of environmental adversity. The diverse responses of the SOD mRNAs suggests that each stress condition produces its own unique effects (Tsang *et al.*, 1991). In *Vitis* protoplasts, SOD activity increased significantly during culture, whereas in *Nicotiana* protoplasts it remained low (Fig. 7). The cytoplasmic SOD isoforms decreased in grapevine during culture and they were not detected after 96h but in tobacco they were pronounced. The results of the activity staining of SOD isoforms after native PAGE and the quantity of the immunoreactive protein detected with specific anti-SOD were in complete accordance (Siminis and Roubelakis-Angelakis, manuscript in preparation).

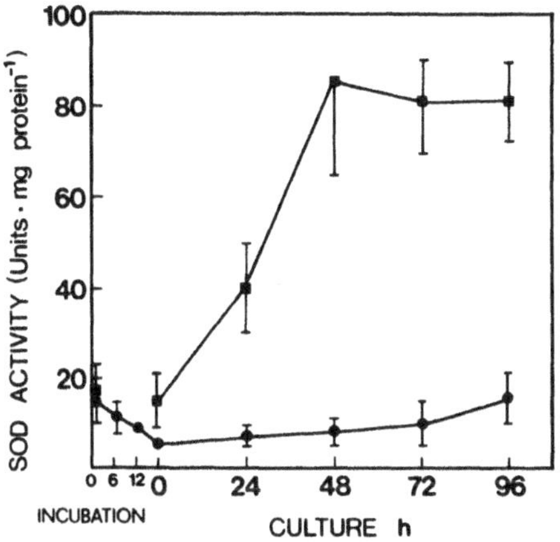

Figure 7. Specific activity of SOD in cultured *Nicotiana tabacum* (■) and *Vitis vinifera* (●) protoplasts (Siminis and Roubelakis-Angelakis, unpublished).

Furthermore during callogenesis from *Vitis* nodal and internodal explants, both catalase and SOD specific activities increased the 3rd week of culture (first visible signs of callogenesis). These increases coincided with a decrease in fluorescent oxidation products, suggesting that the explants are capable of ameliorating oxidative stress induced during early dedifferentiation (Benson and Roubelakis-Angelakis, 1992). Although there are some

Brenneman F, Galston AW (1975) Experiments on the cultivation of agriculturally important plants. I. Oat (*Avena sativa* L). *Biochem Physiol Pflanzen* 168:453-471.

Brown RMJr, Haigler CH, Syttie J, White AR, Oberts E, Smith C, Itoh T, Cooperk (1983) The biosynthesis and degradation of cellulose. *J Appl Polymer Sci* 37:33-78.

Burgess J, Fleming EN (1974) Ultrastructural observations of cell wall regeneration around isolated tobacco protoplasts. *J Cell Sci* 14:439-449.

Burgess J, Linstead PJ (1976) Scanning electron microscopy of cell wall formation around isolated plant protoplasts. *Planta* 131:173-178.

Burgess J, Linstead PJ, Bonsall VE (1978) Observations on the time course of wall development at the surface of isolated protoplasts. *Planta* 139:85-91.

Cocking EC (1961) Properties of isolated protoplasts. *Nature* 191:780-782

Criqui MC, Durr A, Parmentier J, Marbach J, Fleck J, Jamet E (1993) How are photosynthetic genes repressed in freshly-isolated mesophyll protoplasts of *Nicotiana tabacum*. *Plant Physiol Biochem*, in press.

Darken MA (1961) Applications of fluorescent brighteness in biological techniques. *Science* 133:1704-1705.

Donats H (1981) Lipid peroxidation, crosslinking reactions and aging. *In*: Age Pigments, Sohal RS (ed), Elsevier/North-Holland Bio-Medical Press, Amsterdam, pp 63-81.

Eriksson T, Jonasson, K (1969) Isolation and division of *Haplopappus* protoplasts. *Planta* 89:85-90.

Fosket DE, Tonoike H, Colon A (1993) What is the significance of the relatively large tubulin multigene families for plant morphogenesis. *In*: Morphogenesis in Plants: Molecular Approaches Roubelakis-Angelakis KA, Tran Thanh Van K (eds), Plenum Publ Co, New York, pp 55-87.

Frearson EM, Power JP, Cocking EC (1973) The isolation, culture and regeneration of *Petunia* leaf protoplasts. *Developm Biol* 33:130-137.

Fridovich I (1986) Biological effects of the superoxide radical. *Arch Biochem Biophys* 247:1-11.

Galbraith AW, Shields BA (1982) The effects of inhibitors of cell wall synthesis on tobacco protoplast develompment. *Physiol Plant* 55:25-30

Galston AW, Kaur-Sawhney R (1986) Polyamines and senescence in plants. *In*: Plant Senescence: Its Biochemistry and Physiology, Thomson WW, Nothnagel EA, Huffaker RC (eds), *Amer Soc Plant Physiol*, Rockville, MD, USA, pp 167-181.

Giaja F (1919) Emploi des ferments dans les etudes de physiologie cellulaire le globure de levure depouille de sa membrane. *CR Soc Biol Fil Paris* 82:719-723.

Grosset J, Meyer Y, Chartier Y, Kauffmann S, Legrand M, Friting B (1990) Tobacco mesophyll protoplasts synthesize 1,3-glucanase, chitinases, and osmotins during *in vitro* culture. *Plant Physiol* 92:520-527.

Hahne G, Hoffmann F (1984) Dimethyl sulfoxide can initiate cell divisions of arrested callus protoplasts by promoting cortical microtubule assembly. *Proc Natl Acad Sci USA* 81:5449-5453.

Hanstein J (1880) Biologie des protoplasmas. *Bot Abh* 4:1-5.

Herburn AG, Belanger FC, Mattheis, JR (1987) DNA methylation in plants. *Dev Genet* 8:475-493.

Hesemann CU, Schroder G (1982) Loss of nuclear DNA in leaves of rye. *Theor Appl Genet* 62:325-328.

Huhtinen O, Delmer DP, Simola LK (1982/83) Ornithine and putrescine supported divisions and cell colony formation in leaf protoplasts of alder (*Alnus glutinosa* and *A. incana*). *Plant Sci Lett* 29:3-9.

Ishii S (1988) Factors influencing protoplast viability of suspension-cultured rice cells during isolation process. *Plant Physiol* 88:26-29.

Izhar S, Power JB (1977) Genetical studies with *Petunia* leaf protoplasts. I. Genetic variation to specific growth hormones and possible genetic control on stages of protoplast development in culture. *Plant Sci Lett* 8:375-383.

Katsirdakis KC (1991) Cytological, Biochemical and Ultrastructural Studies of Cultured Protoplasts from *Vitis vinifera* L. Ph.D. Dissertation, University of Crete, Heraklio, Greece.

Katsirdakis KC, Roubelakis-Angelakis KA (1992a) A modified culture medium and culture conditions increase viability and cell wall synthesis in grapevine (*Vitis vinifera* L cv Sultanina) leaf protoplasts. *Plant Cell Tiss Org Cult* 28:255-260.

Katsirdakis KC, Roubelakis-Angelakis KA (1992b) Ultrastructural and biochemical aspects of cell wall regeneration in recalcitrant and regenerating leaf protoplasts. *In vitro Cell Dev Biol* 28:90-97.

Klee CB, Vanaman TC (1982) Calmodulin. *Adv Prot Chem* 35:213-321.

Klercker I (1892) Eine methode zur isolierung lebender protoplasten. *Ofvers vet-Akad Forhdl* 9:463-474.

Koornneef M, Hanhart CL, Martinelli L (1987) A genetic analysis of cell culture traits in tomato. *Theor Appl Genet* 74:633-641.

Lett MC, Fleck J, Fritsch C, Durr A, Hirth, L (1980) Suitable conditions for characterization, identification and isolation of the mRNA of the small subunit of ribulose-1,5-biphosphate carboxylase from *Nicotiana sylvestris*. *Planta* 148:211-216.

Ling V, Assmann SM (1992) Cellular distribution of calmodulin and calmodulin-binding proteins in *Vicia faba* L. *Plant Physiol* 100:970-978.

Maheshwari SC, Gill R, Maheshwari N, Gharyal RK (1986) Isolation and regeneration of protoplasts form higher plants. *In*: Differentiation of Protoplasts and of Tranformed Plant Cells, Reinert J, Binding H (eds), Springer-Verlag, Berlin, pp 3-36.

McClelland M, Nelson M (1985) The effect of site specific methylation on restriction endonuclease digestion. *Nucleic Acids Res* 13:r201-r207.

Meins FM (1983) Heritable variation in plant cell cultures. *Annu Rev Plant Physiol* 34:327-346.

Meyer Y, Herth W (1978) Chemical inhibition of cell wall formation and cytokinesis, but not nuclear division, in protoplasts of *Nicotiana tabacum* cultivated *in vitro*. *Planta* 142:253-262.

Meyer Y, Chartier Y, Grosset J, Marty I, Brugidou C, Marinho P, Rivera R (1993) Gene expression in mesophyll protoplasts. *In*: Morphogenesis in Plants: Molecular Approaches, Roubelakis-Angelakis KA, Tran Thanh Van K (eds), Plenum Publ Co, New York, pp 221-236.

Mock H-P, Emmerling M, Seitz HU (1990) Cell wall synthesis in carrot cells: Comparison of suspension-cultured cells and regenerating protoplasts. *Physiol Plant* 79:347-353.

Morrish FM, Vasil IK (1989) DNA methylation and embryogenic competence in leaves and callus of Napiergrass (*Pennisetum purpureum Schum*). *Plant Physiol* 90:37-40.

Nagata T, Takebe I (1970) Cell wall regeneration and cell division in isolated tobacco mesophyll protoplasts. *Planta* 92:301-308.

Necas O (1980) Regeneration of protoplasts. *In*: Advances in Protoplast Research, Ferenczy L, Farkas GL (eds), Pergamon Press, Oxford, pp 151-161.

Nick H, Bowen B, Ferl RJ, Gilbert W (1986) Detection of cytosine methylation in the maize alcohol dehydrogenase gene by genomic sequencing. *Nature* 319:243-246.

Osborn M, Weber K (1980) Dimethyl sulfoxide and the ionophore A 23187 affect the arrangement of actin and induce nuclear actin paracrystals in ptk2 cells. *Exp Cell Res* 129:103-114.

Pearce RS, Cocking EC (1973) Behaviour in culture of isolated protoplasts from "Paul's Scarlet" rose suspension culture cells. *Protoplasma* 77:165-170.

Perera IY, Zielinski RE (1992) Synthesis and accumulation of calmodulin in suspension cultures of carrot (*Daucus carota* L.). *Plant Physiol* 100:812-819.

Roberts RJ (1980) Restriction and modification enzymes and their recognition sequences. *Nucleic Acids Res* 8:r63-r80.

Roubelakis KA, Kliewer WM (1978a) Enzymes of Krebs-Henseleit Cycle in *Vitis vinifera* L. I. Ornithine transcarbamoylase: isolation and some properties. *Plant Physiol* 62:337-339.

Roubelakis KA, Kliewer WM (1978b) Enzymes of Krebs-Henseleit Cycle in *Vitis vinifera* L. II. Arginosuccinate synthetase and lyase. *Plant Physiol* 62:340-343.

Roubelakis KA, Kliewer WM (1978c) Enzymes of Krebs-Henseleit Cycle in *Vitis vinifera* L. III. *In vivo* and *in vitro* studies of arginase. *Plant Physiol* 62:344-347.

Roubelakis-Angelakis KA, Theodoropoulos PA (1990) Uptake characteristics of sugars and amino acids by *Vitis vinifera* L protoplasts. *In*: Molecular Basis of Plant Aging, Rodriguez R, Tames RS, Durzan D (eds), Plenum Publ Corp, New York, pp 89-96.

Saxena RK, Rashid A (1981) High frequency regeneration of *Funaria hygrometrica* protoplasts isolated from low calcium protonemal suspension. *Plant Sci Lett* 23:117-122.

Seitz MV, Seitz U, Alfermann W (1985) Pflanzliche Gewebekultur. Ein Praktikum Gustav Fischer Verlag, Stuttgart, pp 85-87.

Schwartz D, Dennis E (1986) Transposase activity of the AC controlling element in maize is regulated by its degree of methylation. *Mol Gen Genet* 205:476-482.

Shepard JF (1981) Protoplasts as sources of disease resistance in plants. *Annu Rev Phytopathol* 19:145-166.

Siminis CI, Kanellis AK, Roubelakis-Angelakis KA (1990) *De novo* protein synthesis in tobacco and grapevine protoplasts during culture. *Physiol Plant* 79:483.

Siminis, CI, Kanellis AK, Roubelakis-Angelakis KA (1992) Catalase is differentially expressed in regenerating and recalcitrant protoplasts during isolation and culture. *Physiol Plant* 85:A95.

Siminis CI, Kanellis AK, Roubelakis-Angelakis KA (1993a) Differences in protein synthesis and peroxidase isoenzymes between recalcitrant and regenerating protoplasts. *Physiol Plant*, 87:263-270.

Siminis CI, Kanellis AK, Roubelakis-Angelakis KA (1993b) Different catalase subunits are expressed in recalcitrant and dividing tobacco protoplasts, submitted.

Slocum RD, Kaur-Sawhney R, Galston AW (1984) The physiology and biochemistry of polyamines in plants. *Arch Biochem Biophys* 235:283-303.

Takebe I, Labib G, Melchers G (1971) Regeneration of whole plants form isolated mesophyll protoplasts of tobacco. *Naturwissenschaften* 58:318-320.

Takeuchi Y, Komamine A (1978) Composition of the cell wall formed by protoplasts isolated from cell suspension cultures of *Vinca rosea*. *Planta* 140:227-232.

Takeuchi Y, Komamine A (1982) Effects of culture conditions on cell division and composition of regenerated cell walls in *Vinca rosea* protoplasts. *Plant Cell Physiol* 23:249-255.

Theodoropoulos PA, Roubelakis-Angelakis KA (1989) Mechanism of arginine transport in *Vitis vinifera* L. protoplasts. *J Exp Bot* 40:1123-1230.

Theodoropoulos PA, Roubelakis-Angelakis KA (1990) Progress in isolation and culture of leaf protoplasts from virus- free axenic shoot cultures of *Vitis vinifera* L. *Plant Cell Tiss Org Cult* 20:15-23.

Theodoropoulos PA, Roubelakis-Angelakis KA (1991) Glucose transport in *Vitis vinifera* L. protoplasts. *J Exp Bot* 42:477-483.

Tiburcio A, Campos FL, Figueras X, Marce M, Capell T, Riera R, Besford ET (1993) Polyamines and Morphogenesis in Monocots: Experimental Systems and Mechanisms of Action. *In*: Morphogenesis in Plants: Molecular Approaches, Roubelakis-Angelakis KA, Tran Thanh Van K (eds), Plenum Publ Co, New York, pp 113-135.

Tsang WT Ed, Bowler C, Herouart D, Camp WV, Villarroel R, Genetello C, Van Montagu M, Inze D (1991) Differential regulation of superoxide dismutases in plants exposed to environmetal stress. *Plant Cell* 3:783-792.

Vasil V, Vasil IK (1980) Isolation and culture of cereal protoplasts. 2. Embryogenesis and plantlet formation from protoplasts of *Pennisetum americanum. Theor Appl Genet* 56:97-99.

Vernet T, Fleck J, Durr A, Fritsch C, Pinck M, Hirth L (1982) Expression of the gene coding for the small subunit of ribulose biphosphate carboxylase during differentiation of tabacco plant protoplasts. *Eur J Biochem* 126:489-424.

Wang H, Cutler AJ, Saleem M, Fowke LC (1989) DNA replication in maize leaf protoplasts. *Plant Cell Tiss Org Cult* 18:33-46.

Wang MC, Lin JJ, Duran-Vila N, Semancik IS (1986) Alteration in cell wall composition and structure in viroid-intected cells. *Physiol Mol Plant Path* 28:107-124.

Willison JHM, Groot BW (1978) Further observations on cell wall formation around isolated protoplasts of tobacco and tomato. *Planta* 140:53-58.

greatly modify this structure (Meyer and Herth, 1978). During the first two days of culture, the protoplasts become bigger and the chloroplasts surround the nucleus like a rosette, cytoplasmic strands are formed. At the same time cell wall regeneration starts and cell shape changes from round to oval. The cytoplasmic ribosome and polysome density and the number of dictyosomes increase. Cytoskeleton is profondly modified (Hahne and Hoffmann 1985; Le Guyader, 1990; Meijer *et al.*, 1988). Nuclear division takes place after two days in culture. At the same time chloroplast thylakoids are disrupted and chloroplasts change to amyloplasts. After one week in culture cells derived from mesophyll protoplasts show the morphology of callus cells (Herth and Meyer, 1978).

At the molecular level the very low thymidine incorporation observed at the begining of culture is followed after 30 hours by a massive incorporation (Zelcer and Galun, 1976,1980) and an increase of the DNA content (Galbraith *et al.*, 1981) indicating the entry into S phase. Uridine and methionine incorporation rise during the first hours in culture indicating active RNA (Coutts *et al.*, 1975; Sakai and Takebe, 1970) and protein synthesis (Bergounioux *et al.*, 1988a; Gigot *et al.*, 1975; Cooke and Meyer, 1981; Ruzicska *et al.*, 1979;).

Culture medium for mesophyll protoplasts is similar to a callus culture medium, with the addition of an osmoticum in order to replace the wall pressure (Nagata and Takebe, 1970). In the absence of hormones mesophyll protoplasts do not divide: A kinetic analysis shows that auxin is necessary from the begining of culture for the progression into the cell cycle, while cytokinin intervenes later in culture, just before S phase. In addition to this effect on mitosis, auxin and cytokinin profoundly affect the development of protoplasts in culture: In the absence of one of these hormones cells become bigger, cytoplasmic strands are rare, cell wall reformation is limited and the culture medium become toxic (Bergounioux *et al.*, 1988b; Meyer and Cooke, 1979).

Changes in a few enzymatic activities during protoplast culture have been reported: RNAse activities rise 10 to 15 times, apparently as a consequence of the osmotic pressure (Lazar *et al.*, 1973). Peroxidase activities increase in culture (Mäder *et al.*, 1977; Siminis *et al.*, 1993).

The first published electrophoretic patterns (Fleck *et al.*, 1979, 1980) show that the most abundant leaf proteins are no longer synthesized in protoplasts a few hours after isolation, and that most of the proteins synthesized by protoplasts are undetectable in the leaf. Most of these changes are independent of the composition of the culture medium (Meyer *et al.*, 1984a,b). Thus, at that time it appeared clear that changes in protein synthesis take place early during or after protoplast isolation, and that the isolation has a far more important impact than the composition of the culture medium.

CHANGES IN GENE EXPRESSION

In recent years molecular cloning of plant genes has developed rapidly and the approach to the analysis of early development has shifted from protein synthesis to mRNA analysis.

Photosynthesis Related Genes

Most of the metabolic activity in mesophyll cells is centered on photosynthesis. Both nuclear and chloroplast genes participate in the synthesis of the chloroplast. In particular ribulose 1,5-bisphosphate carboxylase (RuBP) is composed of 8 identical small subunits (SSU) encoded by the nuclear genome, associated with 8 identical large subunits (LSU) encoded by the chloroplast genome. This enzyme, located in the chloroplast stroma, catalyses the first reaction in the Calvin cycle, namely CO_2 fixation on ribulose 1,5-bisphosphate. In addition to RuBP mRNAs, the fate of mRNAs coding for a light harversting chlorophyll a/b binding protein from photosystem II (LHCII) as well as two chloroplast genes encoding a quinone fixing protein (DI) and a protein of photosystem I (P700) have been analysed during protoplastization and in culture. All mRNAs corresponding to these genes are abundant in tobacco leaf. The mRNAs encoded by the nuclear genes decline rapidly during protoplast isolation (Lett *et al.*, 1980; Vernet *et al.*, 1982), while these corresponding to the chloroplast genes continue to be expressed at a high level for at least two days (Criqui *et al.*, 1992a). They decline only later, when the chloroplast structure begin to be altered and starch begins to be accumulated in the plastid, i.e. when the transition from chloroplast to amyloplast is clearly detectable. Absence of coordination between the expression of the nuclear and chloroplast encoded genes has previously been observed during the shift from chloroplast to chromoplast. It is important to add that in spite of the presence in the freshly isolated protoplast of chloroplast LSU mRNAs at a level similar to the leaf, the corresponding polypeptide is not synthesized in the protoplasts, as shown by ^{35}S methionine labeling.

The same development is observed when leaf slices are cultivated in the protoplast medium or in water, indicating that neither plasmolysis nor maceration are necessary for the repression of the photosynthesis related nuclear genes. Sucrose (Criqui *et al.*, 1992a) and auxin (Takahashi *et al.*, 1989), if present in the maceration or culture medium, activate the decline of the mRNA level of the nuclear encoded photosynthesis-related genes. Nevertheless, the phenomenon, although slower, takes place in the absence of any metabolisable sugar or hormone suggesting that the rapid repression of these nuclear genes in mesophyll protoplasts is primarily induced by the mechanical wounding stress applied to the leaf prior to maceration.

Pathogenesis-Related Proteins

Tobacco plants reacting against TMV infection by hypersensitivity actively accumulate a small set of proteins which have been named Pathogenesis-related proteins (PR) (Van Loon, 1985). They are acidic, localized in the cell wall and the enzymatic activity of some of them has been characterized as hydrolases (Legrand *et al.* 1987; Kaufmann *et al.*, 1987) potentially able to destroy the cell wall of pathogens or saprophytes (Mauch *et al.*, 1988). They are generally considered to be defence proteins. More recently, it has been shown that another set of proteins is synthesized after TMV infection. They are basic, located in the vacuole and show considerable homologies with the acidic ones (Hooft van Huijsduijnen *et al.*, 1987). Consequently, they have been integrated into the PR proteins. Nevertheless, both groups are not regulated in the same manner: While acidic PRs are inducible by a limited number of stresses (pathogen infection, salicilate treament), basic PRs are also induced by wounding, salt treatment (King *et al.*, 1988; Singh *et al.*, 1987) and are expressed without stress in different organs of the plant: at a very high level in mature roots and at a lower level in flowers and senescing leaves (Memelink *et al.*, 1990) and in *in vitro* cultivated callus (Mohnen *et al.*, 1985; Shinshi *et al.*, 1987). PR proteins are not detectable in seedlings, root tips or in heathly leaves which have just reached full expansion and are used for the isolation of tobacco protoplasts. The vacuolar basic PR proteins (glucanase, chitinases and osmotines) accumulate rapidly in protoplasts during culture. Nevertheless, acid PR are not detectable, with the exception of PR P and PR Q, two acidic chitinases which accumulate at a low level in protoplasts, and are released in low amounts into the culture medium (Grosset *et al.*, 1990). It is interesting to note that the same PR proteins are accumulated by callus cells and in leaf strips cultivated *in vitro*, i.e. the absence of most acidic PR is not related to the absence of the cell wall in protoplasts. In cells cultivated in liquid medium, PRs accumulate during the stationary phase but are undetectable during the growth phase. These proteins accumulate in extremely high amounts in leaf slices cultivated in water, at a level 5 to 10 time higher than in a medium compatible with callus formation. The presence of the osmoticum has little effect on this accumulation. Amongst medium compounds, auxin strongly reduces the mRNA accumulation of the PR proteins and of the corresponding mRNA, both in leaf slices and protoplasts (Grosset *et al.*, 1990).

Extensin

Extensins are glycoproteins having a very high hydroxyproline content (up to 45% of the aminoacids) Lamport (1965). They are located in the cell wall and seem to play a role both in cell growth and cell defence (Memelink *et al.*, 1987).

Using a carrot probe, Criqui (1992) showed the absence of transcript in the leaf. Protoplasts in culture first accumulate a 1.2 kb transcript and later a 1.4 kb transcript. Both transcripts are induced in leaf slices cultivated in the protoplast medium, indicating that the maceration is not responsible for the mRNA accumulation. In addition, these messenger RNAs accumulate to far higher levels when the leaf slices are cultivated in water, indicating that the high osmotic pressure of the culture medium is not necessary for the induction and that some elements of the culture medium inhibits the mRNA accumulation. Furthermore, both mRNAs are detectable in roots. Thus, both extensin transcripts are regulated in a way very similar to the PR mRNAs.

Ubiquitins

Ubiquitin is a small protein involved in the ATP-dependent selective degradation of cellular proteins (Hershko, 1983), maintenance of chromatin structure, regulation of gene expression, stress response (Fornace *et al.*, 1989), and ribosome biogenesis. Ubiquitin is synthesized as polyubiquitin containing two to seven motifs (Swindle *et al.*, 1988), or as a fusion protein comprising a ribosomal protein as a C-terminal extension of ubiquitin (CEP). The synthesis of this form has been associated with cell division and growth.

Using a human cDNA ubiquitin probe, Jamet *et al.* (1990) demonstrate the presence of at least four transcripts containing an ubiquitin sequence in *N. sylvestris*. Cloning shows that the 0.8 kb transcript encodes a CEP (Genschik *et al.*, 1990). It is very abundantly expressed during the growth phase of tobacco cells and in growing tissues. This transcript is undetectable in the mature leaf, and present at a low level in protoplasts. Three other transcripts (1.3, 1.6, 1.9 kb) encode polyubiquitins. The 1.3 kb transcript is present in stressed and dividing tissues, the 1.6 kb transcript is induced early by protoplast isolation but not by other stresses (heat-shock, $HgCl_2$), while the 1.9 kd transcript is present in both protoplasts and stressed tissues (Criqui, 1991, J. Fleck personnal communication).

HMG-CoA-Reductase and Putative Glutathione Peroxidase

Criqui *et al.* (1993a) prepared a cDNA library from *N. sylvestris* protoplasts isolated by an overnight maceration followed by six hours culture in a medium compatible with protoplast division. They selected two clones corresponding to mRNA not detectable in the leaf, expressed both in protoplasts and in leaf slices cultivated in the protoplast medium.

The first clone shows a very high homology to an *Arabidopsis thaliana* HMG-CoA-reductase (70% identity of the nucleotide sequences). This enzyme catalyses the first reaction of the isoprenoid pathway, from which various key cellular compounds are synthesized (mevalonic acid, sterols, carotenoids...) (Bach, 1987; Yang *et al.*, 1991). Its mRNA is present at a low level in apex, anthers and roots but surprisingly absent from seedlings and young leaves. The absence of this mRNA in these rapidly growing organs suggests that other mRNAs encode HMG-Co-reductases. The messenger accumulates at a high level in protoplasts during culture and is well expressed in cell suspensions. It is also induced in leaf slices cultivated *in vitro* in the protoplast medium. $HgCl_2$ vaporisation on unwounded leaves also induces the mRNA at a high level.

The second clone shows 33% identity with animal seleno glutathione peroxidases, which are considered as key detoxification enzymes able to trap hydrogen and lipid peroxides (Ischii, 1987). Although the enzymatic activity of the encoded protein is not definitly established, due to the relatively low sequence homology, the pattern of mRNA accumulation has been analyzed in details: It is expressed at a high level in seedlings but not in mature organs (leaves, stem, roots). It is expressed very early in fresly isolated protoplasts, and then decline in culture. It is also induced in leaf strips, or in unwounded leaf by $HgCl_2$ vaporisation. This mRNA is also detectable in suspension cells.

Anionic Peroxidase, Serine Protease Inhibitor 1, 6P50

Criqui *et al.* (1993b) isolated from their *N. sylvestris* library three cDNA clones not expressed in the leaf, expressed early in protoplasts and only at a low level in leaf slices cultivated in the protoplast medium.

The first clone encodes one of the multiple tobacco peroxidases. *In vitro* peroxidases are able to oxidize guaiacol or other synthetic substrates in the presence of hydrogen peroxide, but the cellular substrate is generally unknown. They are implicated in auxin degradation, lignin and suberin formation, defence against pathogens and possibly germination and root formation (Lagrimini and Rothstein, 1987; Lagrimini *et al.*, 1990). This mRNA is detectable in roots but not in other organs.

sized in the mature leaf. In fact it appears that ribosome synthesis in plants is coupled to cellular proliferation and is not detectable in mature tissues (Lebrun and Freyssinet, 1991), even those which actively synthesize proteins, like seeds during storage protein accumulation (Marty and Meyer, 1992).

A further clone encodes a cyclophilin, a peptidyl-prolyl cis-trans isomerase thus able to modify the conformation of target proteins (Bachinger, 1987). Cyclophilins are associated with cell proliferation in human t lymphocytes (Fischer *et al.*, 1989) and detected in growing plant organs (Gasser *et al.*, 1990; Marivet *et al.*, 1992).

A fifth clone presents no homology with the protein data banks. All five mRNAs are abundant in rapidly-growing tissues, but not in mature organs. They are detectable in protoplasts after 4 hours in culture, or in the maceration solution if the treatment is applied for a longer time. In contrast, if freshly isolated protoplasts are cultivated in KCl+$MgSO_4$, the mRNAs do not become detectable. Thus it appears that something common to the maceration medium and the culture medium is necessary for the induction of these genes. In contrast to thioredoxin and to the homolog of the Wilm's tumor suppressor, sucrose is not necessary for this induction (Marty *et al.*, 1993). No mRNA corresponding to the sixth clone can be detected in protoplasts.

Histone H3 and H4

Histone H3 (Takahashi *et al.*, 1989) and H4 (Marty *et al.*, 1992) probes (Chaboute *et al.*, 1987) have been used as markers of the S phase. In both cases Northern blots with protoplast mRNAs present very low signals even with protoplasts actively incorporating ^{35}S methionine.

DISCUSSION

Recently our knowledge on genes expressed early after mesophyll protoplast isolation have increased significantly, revealing the expression, among others, of some genes which were not suspected to be present in plants some years ago. In fact this situation results from progress in cDNA cloning, sequencing and data base analysis, and is also common in other fields of plant physiology. It is to be hoped that more mRNA transcripts expressed very early will be characterized because this constitutes an important basis for the understanding of the mechanisms and inducers implicated in dedifferentiation. In spite of the relatively limited number of mRNAs presently characterized, the nature of the proteins that they encode deserves some comments: Clearly no gene directly related to mitosis has been shown to be expressed in protoplasts, although plant cDNAs encoding cdc2 (Feiler and Jacobs, 1991; Ferreira *et al.*, 1991) and cyclins (Hemerly *et al.*, 1992) have been isolated in different labs. In fact H3 and H4 histones mRNAs are present only at a low level in mesophyll protoplasts.

There is no doubt that genes controlling mitosis or encoding structural proteins necessary for mitosis (histone, DNA polymerases...) have to be expressed in protoplasts for the progression through the cell cycle, but we can recall that the first cell cycle associated with dedifferentiation is very long. In a population of tobacco mesophyll protoplasts, the first cells divide after 48 hours, but most protoplasts undergo their first mitosis far later, only after 4 to 5 days. This should be compared to the 20-24 hour cell cycle of meristematic or suspension cells. Consequently, a low level of mRNAs encoding mitosis related-proteins is probably sufficient. Even the auxin dependent gene parA is clearly not directly related to mitosis: it is not expressed at a high level in shoot tips and young leaves, organs in which cells actively divide. This is coherent with the pleiotropic effect of auxin which is not only necessary for the progression into the cell cycle, but also plays roles in cell wall regeneration, vacuolisation, cytoplasmic activity and detoxification of the medium. The growth-related genes reported in this review are overexpressed in protoplasts, when compared with proliferating cells. This clearly shows that they are not directly implicated in cell cycle progression. In fact most genes expressed during proliferation are not implicated in mitosis, but produce basal cellular compounds: this is clearly illustrated by the active expression of ribosomal protein and elongation factor genes.

Amongst the twelve genes presently characterized three, selected by independent screenings are related to redox reactions. Although this may be fortuitous, it is also possible that this indicates that this relatively neglected aspect of cell metabolism plays a key role in cell development both in regulating protein activity (thioredoxin) and trapping peroxide radicals (glutathione S-transferase, glutathione peroxidase) produced by the active cell metabolism.

One of the long term goals is to characterize the mediators which induce the mesophyll cell to dedifferentiate. The first step is to determine the events in protoplast isolation which lead to the synthesis of new mRNAs: PR protein and extensin mRNAs are induced in leaf slices (1-2 mm wide) at approximately the same level as in protoplasts, but not or only at a low level in large leaf explants. This is coherent with the participation of an inducer liberated by wounding which migrates approximately one mm in the mesophyll, i.e. 20-30 cell layers. In addition, these mRNAs are far more abundant when leaf slices are maintained in a medium incompatible with the induction of mitosis in the explant. Growth-related mRNAs are induced at a level 2-5 times lower in leaf slices than in protoplasts. This is compatible with an induction in cells near the wound, up to 1 to 2 cell layers away. These genes are not induced in a medium incompatible with growth. A screening has been directed to the characterization of mRNAs mostly abundant in protoplasts, but not in leaf slices cultivated in the protoplast medium. Using the resulting clones, it has later been shown that these mRNAs are already induced in the leaf slices, but at a far lower level than in protoplasts. It is well known that when leaf slices are cultivated, callus formation takes place at the level of the wound and begins in a few spots. Thus this difference in the mRNA levels is compatible with the induction of these genes in a few cells at the wounded surface of the explant. Clearly *in situ* hybridization studies are needed to confirm this interpretation. Nevertheless, wounding has been found to be the inducing step for most of the genes expressed early in protoplasts, although additional conditions in the composition of the culture medium are required for some of them. The induction of Tnt1 is one exception, as it has been attributed to the maceration (Pouteau *et al.*, 1991). In fact it is likely that, due to the particular situation of protoplasts, not all genes expressed early in these cells are related to dedifferentiation, and that some of them may results from the high osmotic pressure of the culture medium (Fleck *et al.*, 1982), from the use of crude enzymes or from the maceration.

Nevertheless, the studies on large leaf explants and slices very clearly show that most genes expressed early in protoplasts are induced by wounding and that the original wound signal is probably relayed by secondary messangers which present different mobilities in plant tissues. Wound-induced genes have been described (Hilmann *et al.*, 1992; Pena-Cortes *et al.*, 1988; Stanford *et al.*, 1990; Weissand and Bevan, 1991) and for some of them the promoter has been analyzed (Logemann *et al.*, 1989). However, mediators have been characterized only in the case of proteinase inhibitor II. After wounding of tomato plants proteinase inhibitors II are synthesized in a very systemic way: wounding one leaf induces the synthesis of proteinase inhibitor II in other leaves (Ryan, 1987). It was first shown that oligosaccharide wall compounds are released by wounding and have the potential to induce the systemic response, but that they do not migrate into the plant, suggesting the induction of a second messenger (Ryan and An, 1988). This has been recently shown to be a protein, which has been named systemin. Thus, this model predicts that wounding releases non-migrating oligosaccharides which induce systemin gene expression, systemin synthesis and secretion in cells near the wound (Pearche *et al.*, 1991). In addition, another compound, jasmonic acid, seems to be a further intermediate in the induction of proteinases inhibitors II in the target cells (Farmer and Ryan, 1992). This illustrates very well how complex a plant response to a particular effector can be. The complexity and diversity of the reactions of plants to wounding can probably be better understood from a physiological and ecological point of view. Plants do not have the possibility to move to protect themself against aggressions, but possess different defence strategies which are complementary. Wounding frequently results from insect feeding. After a first injury some plants resist a possible invasion by systemic synthesis of proteinase inhibitors that make the rest of the plant inedible. In addition, the open wound is a very favourable site for pathogens and even for saprophyte entry. Plants react by a local or nearly local induction of defence proteins which block the growth of parasites or by necrosis, the dead cells constituting a medium which is not adapted to the growth of parasites. Finally, if the physiological condition of the tissues is adequate, some cells recover their proliferation potential, form a wound callus which repairs the damage, even when this is considerable (i.e. grafting), or form new meristems if necessary (cuttings, decapitated plants). All the genes playing a role in these different processes could be described as wound-inducible, but are nevertheless regulated in very different ways and probably have no common structures in their promoters.

Our present knowledge of gene expression in mesophyll protoplasts shows that at least defence and proliferation-related genes are expressed early and continue to be expressed in *in vitro* callus. This shows the strong similarity between *in planta* healing callus and *in vitro* cultivated callus. In addition to common gene expression, *in vitro* callus has most of the characteristics of healing callus, including the ability to form meristems and there is a striking parallel between the ability of some tissues to form a healing callus after wounding and to proliferate *in vitro* (very high potential of the cambium of dicots, extremely low potential of the mature tissues of cereals). Thus, we propose that the analysis of early gene expression in mesophyll protoplasts will provide information relevant to general processes of plant development, including dedifferentiation and wound reactions, and possibly provide solutions which will help to overcome the inability (Siminis *et al.*, 1992) of some cells to dedifferentiate and form callus *in vitro* for example pollen or mature cells in cereals (Hahne *et al.*, 1989; Wang *et al.*, 1989).

ACKNWOLEDGEMENTS

Authors thank Dr Richard Cooke (Perpignan) for his help in the sequence analysis and in the english translation, and Dr Jacqueline Fleck (Strasbourg) and Dr Marie-Angele Grandbastien (Versailles) for the communication of results before publication and interesting discussions.

REFERENCES

Bach TJ (1987) Synthesis and metabolism of mevalonic acid in plants. *Plant Physiol Biochem* 25: 163-178

Bachinger HP (1987) The infuence of peptidyl-prolyl cis-trans isomerase on *in vitro* folding of type III collagen. *J Biol Chem* 262: 17144-17148.

Bergounioux C, Perennes C, Brown SC, Gadal P (1988a) Nuclear RNA quantification in protoplast cell-cycle phases. *Cytometry* 9: 84-87.

Bergounioux C, Perennes C, Brown SC, Gadal P (1988b) Cytometric analysis of growth regulator-dependent transcription and cell cycle progression in *Petunia* protoplast cultures. *Planta* 175: 500-505.

Brugidou C, Marty I, Chartier Y, Meyer Y (1993) The *Nicotiana tabacum* genome encodes two cytoplasmic thioredoxin gene which are differently expressed. *Mol Gen Gen*, in press.

Chaboute ME, Chaubet N, Philipps G, Gigot G (1987) Genomic organization and nucleotide sequences of two histone H3 and two histones H4 genes of *Arabidopsis thaliana*. *Plant Mol Biol* 8:179-191.

Cooke R, Meyer Y (1981) Hormonal control of tobacco protoplast nucleic acid metabolism during *in vitro* culture. *Planta* 52: 1-7.

Coutts RHA, Barnett A, Wood LR (1975) Ribosomal RNA metabolism in cucumber leaf mesophyll protoplasts. *Nucleic Acid Res* 2: 1111-1121.

Criqui MC (1992) Evènements moleculaires prècoces dans les protoplastes issus de cellules de mèsophylle de *Nicotiana sylvestris* Thése Universite Louis Pasteur - Strasbourg.

Criqui MC, Jamet E, Parmentier Y, Marbach J, Durr A, Fleck J (1991a) Isolation and characterization of a plant cDNA showing homologies to animal glutathione peroxidases. *Plant Mol Biol* 18: 623-627.

Criqui MC, Genschik P, Marbach J, Parmentier Y, Durr A, Bertolino E, Jamet E, Fleck J (1991b) L'ubiquitine joue-t-elle un role dans l'expression de la totipotence de la cellule vegetale? *In*: Chromatine, Aspects Structuraux et Fonctionels, Marion C, J Roche (eds), Edition du CNRS, pp 241-248.

Criqui MC, Durr A, Parmentier J, Marbach J, Fleck JJamet E (1993a) How photosynthetic genes are repressed in freshly-isolated mesophyll protoplasts of *Nicotiana tabacum*. *Plant Physiol and Biochem*, in press.

Criqui MC, Plesse B, Durr A, Marbach J, Parmentier Y, Fleck J (1993b) Characterization of genes expressed in mesophyll protoplasts of *Nicotiana sylvestris* before the re-initiation of the DNA replicational activity. *Mechanisms of development*, in press.

Dominov JA, Stenzler L, Lee S, Schwarz JJ, Leisner S Howell SH (1992) Cytokinins and auxins control the expression of a gene in *Nicotiana plumbaginifolia* cells by feedback regulation. *The Plant Cell* 4: 451-461.

Mäder M, Meyer Y, Bopp M (1976) Zellwandregeneration und peroxidase-isoenzym-synthese isolierter protoplasten von *Nicotiana tabacum* L. *Planta* 129: 33-38.

Marivet J, Frendo P, Burkard G (1992) Effects of abiotic stresses on cyclophilin gene expression in maize and bean and, sequence analysis of bean cyclophilin cDNA. *Plant Science* 84:171-178.

Marty I, Meyer Y (1991) Nucleotide sequence of a cDNA encoding a tobacco thioredoxin. *Plant Mol Biol* 17: 143-147.

Marty I, Meyer Y (1992) cDNA nucleotide sequence and expression of a tobacco cytoplasmic ribosomal protein L2 gene. *Nucleic Acid Res* 20: 1517-1522.

Marty I, Brugidou C, Chartier Y (1993) Growth-related gene expression in *Nicotiana tabacum* mesophyll protoplasts. *The Plant Journal*, in press.

Mauch F, Mauch-Mani B, Boller T (1988) Antifungal hydrolase in pea tissue: Inhibition of fungal growth by combinations of chitinase and b-1,3 -glucanase. *Plant Physiol* 88:936-942.

Meijer EG, Keller WA, Simmonds DH (1988) Microtubule organization during the development of the mitotic apparatus in cultivated mesophyll protoplasts of higher plants - an immunofluorescence microscopic study. *Physiol Plant* 74: 225-232.

Memlink J, Hoge JHC, Schilperoort RA (1987) Cytokinin stress changes the developmental regulation of several defense-related genes in tobacco. *EMBO J* 6: 3579-3583.

Memelink J, Linthorst H, Schilperoort R, Hoge H (1990) Tobacco genes encoding acidic and basis isoforms of pathogenesis-related proteins display different expression patterns. *Plant Mol Biol* 14: 119-126.

Meyer Y, Herth W (1978) Chemical inhibition of cell wall formation and cytokinesis, but not of nuclear division in protoplasts of *Nicotiana tabacum* L cultivated *in vitro*. *Planta* 142: 253-262.

Meyer Y, Cooke R (1979) Time course of hormonal control of the first mitosis in tobacco mesophyll protoplasts cultivated *in vitro*. *Planta* 147: 181-185.

Meyer Y, Aspart L, Chartier Y (1984a) Auxin-induced regulation of protein synthesis in tobacco mesophyll protoplast cultivated *in vitro* : 1. Characteristics of auxin-sensitive proteins. *Plant Physiol* 75: 1027-1033.

Meyer Y, Aspart L, Chartier Y (1984b) Auxin-induced regulation of protein synthesis in tobacco mesophyll protoplast cultivated *in vitro* : 2. Time course and level of auxin control. *Plant Physiol* 75: 51034-1039.

Mohnen D, Shinshi H, Felix G, Meins FJR (1985) Hormonal regulation of β-1,3-glucanase messenger RNA level in cultivated tobacco tissues. *EMBO J* 4:1631-1635.

MBller EGD (1991) Thioredoxin deficiency in yeast prolongs S phase and shortens the G1 interval of the cell cycle. *J Biol Chem* 266: 9194-9202.

Nagata T, Takebe I (1970) Cell wall regeneration and cell division in isolated tobacco mesophyll protoplasts. *Planta* 92: 301-308.

Pearce G, Strydom D, Johnson C, Ryan CA (1991) A polypeptide from tomato leaves induces wound-inducible proteinase inhibitor proteins. *Science* 253: 895-898.

Pena-Cortes H, Sanchez-Serrano J, Rocha Sosa M, Willmitzer L (1988) Systemic induction of proteinase-inhibitor II gene expression in potato plants by wounding. *Planta* 174: 84-89.

Pokalsky AR, Hiatt WR, Ridge N, Rasmussen R, Houk CM, Shewmaker CK (1989) Structure and expression of elongation factor- 1a in tomato. *Nucleic Acids Res* 17: 4661-4673.

Pouteau S, Huttner E, Grandbastien MA, Caboche M (1991) Specific expression of the tobacco Tnt1 retrotransposon in protoplasts. *EMBO J* 10: 1911-1918.

Rivera-Madrid R, Marinho P, Chartier Y, Meyer Y (1993) Nucleotide sequence of an *Arabidopsis thaliana* cDNA clone encoding a homologue to a suppressor of Wilm's Tumor. *Plant Physiol*, in press.

Ruzicska P, Mettrie R, Dorokhov YL, Premecz G, Olah T, Farkas GL (1979) Polyribosomes in protoplasts isolated from tobacco leaves. *Planta* 145; 199-203.

Ryans CA (1987) Oligosaccharide signalling in plants. *Annu Rev Cell Biol* 3: 295-31.

Ryans CA (1989) Proteinase inhibitor gene families: strategies for transformation to improve plant defenses against herbivore. *Bio Essays* 10: 20-24.

Ryans CA, An G (1988) Molecular biology of wound-inducible proteinase inhibitors in plants. *Plant Cell Envir* 11: 345-349.

Sakai F, Takebe I (1970) RNA and protein synthesis in protoplasts isolated from tobacco leaves. *Biochim Biophys Acta* 224: 531-540.

Shinshi H, Mohnen D, Meins FJR (1987) Regulation of a plant pathogenesis related enzyme: inhibition of chitinase and chitinase mRNA accumulation in cultured tobacco tissues by auxin and cytokinin. *Proc Natl Acad Sci* 84: 89-93.

Siminis CI, Kanellis AK, Roubelakis-Angelakis KA (1993) Differences in protein synthesis and peroxidase isoenzymes between recalcitrant and regenerating protoplasts. *Physiol Plant*, 87:263-270.

Singh NK, Bracker CA, Hasegawa PM, Handa AK, Bulckel S, Hermodson MA, Plankoch ED, Regnier FE, Bressan RA (1987) Characterization of osmotin: a thaumatin-like protein associated with osmotic adaptation in plant cells. *Plant Physiol* 85: 529-536.

Stanford AC, Northcote DC, Bevan MW (1990) Spatial and temporal patterns of transcription of a wound-induced gene in potato. *EMBO J* 9: 200-208.

Suthanthiran M, Anderson M, Sharma VK, Meister A (1990) Glutathione regulates activation-dependant DNA synthesis in highly purified normal human T lymphocytes stimulated via the CD2 and CD3 antigens. *Proc Natl Acad Sci USA* 87: 3343-3347.

Swindle J, Ajioka H, Eisen H, Sanwal B, Jacquemont Z, Browder Z, Buck G (1988) The genomic organization and transcription of the ubiquitin genes of trypanosoma cruzi. *EMBO J* 7: 1121-1127.

Tagaya Y, Maeda Y, Mitsui A, Kondo N, Matsui H, Hamuro J, Brown N, Arai K, Yokota T, Wakasugi H, Yodoi J (1989) ATL-derived factor (Adf), an Il-2 receptor TaC inducer homologous to thioredoxin: possible involvement of dithiol reduction in the Il-2 receptor induction. *EMBO J* 8 : 757-764.

Takahashi N, Hayano T, Suzuki M (1989) Peptidyl-prolyl cis-trans isomerase is the cyclosporin A-binding protein cyclophilin. *Nature* 337: 473-475.

Takahashi Y, Kuroda H, Tanaka T, Machida Y, Takebe I, Nagata T (1989) Isolation of an auxin-regulated gene cDNA expressed during the transition from G_0 to S phase in tobacco mesophyll protoplasts. *Proc Natl Acad Sci USA* 86: 9279-9283.

Takahashi Y, Nagata T (1992) parB: an auxin-regulated gene encoding glutathione S-transferase. *Proc Natl Acad Sci* 89: 56-59.

Ursin VM, Irvine JM, Hiatt WR, Shewmaker CK (1991) Developmental analysis of elongation factor-1a expression in transgenic tobacco. *Plant Cell* 3: 583-591.

Van Loon LC (1985) Pathogenesis-related proteins. *Plant Mol Biol* 4: 111-116.

Vernet T, Fleck J, Durr A, Fritsch C, Pinck M, Hirth L (1982) Expression of the gene coding for the small subunit of ribulosebiphosphatase carboxylase during differentiation of tobacco plant protoplasts. *Eur J Biochem* 126: 489-494.

Wang H, Cutler AJ, Saleem M, Fowke LC (1989) DNA replication in maize leaf protoplasts. *Plant Cell Tiss Org Cult* 18: 33-46.

Weiss C, Bevan M (1991) Ethylene and a wound signal modulate local and systemic transcription of win2 genes in transgenic potato plants. *Plant Physiol* 96: 943-951.

Wolosiuk RA, Buchanan BB (1977) Thioredoxin and glutathione regulate photosynthesis in chloroplasts. *Nature* 266: 565-567.

Yang Z, Park H, Lacy GH, Cramer CL (1991) Differential activation of potato 3-hydroxy-methylglutaryl coenzyme A reductase genes by wounding and pathogen challenge. *Plant Cell* 3: 397-405.

Zelcer A, Galun E (1976) Culture of newly isolated tobacco protoplasts: precursor incorporation into protein, RNA and DNA. *Plant Sci Lett* 7: 331-336.

Zelcer A, Galun E (1980) Culture of newly isolated tobacco protoplasts: Cell division and precursor incorporation following a transient exposure to coumarin. *Plant Sci Lett* 18: 185-190.

ANTISENSE RNA IN PLANTS: A TOOL FOR ANALYSIS AND SUPPRESSION OF GENE FUNCTION

Martin Tabler

Institute of Molecular Biology and Biotechnology
Foundation for Research and Technology
P.O. Box 1527, 711 10 Heraklion, GREECE

INTRODUCTION

Expression of antisense genes has become a widely applied technique in plant biotechnology for creating "phenotypic mutants" in which the endogenous target gene is specifically suppressed. The antisense technology is used for various purposes such as silencing or ablating undesired genes. By creating "dominant negative mutants", gene function can also be assigned to a particular cloned cDNA. Despite the rapidly increasing number of successful applications of antisense genes in plants, the molecular mechanisms in which complementary RNAs down-regulate gene expression is not fully understood.

This presentation will give an update on various examples of the expression of antisense RNAs in plants and will outline some characteristics of the technology. It also will discuss some aspects for potential further improvements of the technique by incorporating ribozyme domains into the antisense RNA. Additional information on some aspects of expressing antisense RNAs in plants can be found in van der Krol *et al.* (1988b,c) and Rothstein and Lagrimini (1989).

PRINCIPLE OF THE ANTISENSE TECHNOLOGY

The principle of the antisense technology is based on the capability of complementary nucleic acids to form double-stranded helices. Any cellular single-stranded nucleic acid is therefore susceptible to bind *via* Watson-Crick base pairing to a nucleic acid of complementary (antisense) polarity. Cellular DNA usually is double-stranded, having just a partially single-stranded phase during replication and thus is not a very good target for an antisense nucleic acid. Cellular RNA normally is present only as a single-stranded molecule even if it has double-stranded domains originating from the intramolecular secondary structure. Due to its single-stranded nature, the endogenous RNA is the preferred target for antisense

Morphogenesis in Plants, Edited by K.A. Roubelakis-Angelakis
and K. Tran Thanh Van, Plenum Press, New York, 1993

Figure 1. Map of a typical antisence gene. Antisense RNA is expressed from a DNA cassette that consists of a strong promoter, usually the promoter of the 35S transcript of cauliflower mosaic (CaMV). Downstream of the promoter is the cDNA, or a fragment of it, inserted in antisense orientation which is followed by a transcription terminator that is in many cases derived from the nopaline synthase gene (*nos*).

consists of a strong promoter, usually the promoter of the 35S transcript of cauliflower mosaic virus (CaMV), followed downstream by the cDNA or a fragment thereof, in antisense orientation and a transcription terminator which in many cases is derived from the nopaline synthase gene (*nos*). The antisense technology so far has been applied in seven different plant species: Alfalfa, carrot, petunia, rape, potato, tobacco and tomato, which are a good representation of species that are relatively easy to transform. There is no particular reason to assume that the technique would not work also in other plant species. The genes that have been targeted can be subdivided into two classes. The first class are non-plant genes which had been introduced by stable transformation in order to test the applicability of gene suppression by antisense RNA. Such bacterial or fungal "markers" are chloramphenicol acetyltransferase (CAT), α-glucuronidase (GUS), phosphinotricin acetyltransferase (PAT) or nopaline synthase (NOS). The second class of genes comprise endogenous genes which are involved in such diverse biological processes as flower pigmentation, fruit ripening, photosynthesis, starch biosynthesis, defense response or fatty acid metabolism.

Rather than summarizing the results of the individual reports separately, it is intended to focus on some characteristics common to all the antisense experiments and to outline some specific issues separately.

EXPRESSION OF ANTISENSE RNAs AGAINST MARKER GENES

The expression of antisense RNA that is directed against marker genes requires double-transformation, first for the marker gene and second for the antisense gene. In view of the recent results of Matzke *et al.* (1989) (compare also Hobbs *et al.*, 1990; Matzke and Matzke, 1991) who report on negative interference of two sequentially transformed DNAs due to changes in the methylation pattern it, in retrospect, cannot be excluded that the observed down-regulation of these bacterial or fungal marker genes in part is also caused by other mechanisms than by antisense RNA. In contrast to marker genes, endogenous genes require just one transformation event so that this phenomenon can be disregarded.

EXPRESSION OF AN ANTISENSE GENE RESULTS IN DIFFERENT PHENOTYPES IN INDIVIDUAL PLANTS

The first approach to target an endogenous plant gene by expression of antisense RNA was reported by van der Krol *et al.* (1988a, compare Table I, No 8). They targeted the gene encoding chalcone synthase (CHS) of petunia. CHS is the key enzyme for the biosynthesis of flavonoids which are involved in several biological processes including flower pigmentation. One advantage in targeting this gene is that the resulting phenotype is readily scorable by eye and does not necessitate detailed biochemical analysis. Van der Krol *et al.* (1988a) have reported the occurrence of different classes of transgenic plants that expressed CHS-directed antisense RNA. One class did not show any visible effect as compared to the control plants despite the presence of the antisense RNA. In a second class of transformed petunia plants, flower pigmentation was reduced in the corolla often with white sectors, the tube however, remaining colored. In the third class, color formation started in the tube and extended outward into the corolla tissue, leading either to white flowers with a colored ring or entirely white flowers. In an ensuing report, van der Krol *et al.* (1990a) describe that mixed phenotypes can vary in an individual plant. The phenotype can be either stable or unstable and it can be variable depending on physiological conditions like changes of light condition, hormone concentration or water supply.

Such variability in the "antisense phenotype" has been reported for all the endogenous genes that have been targeted to date. Stockhaus *et al.* (1990, Table I, No 15), who had targeted the mRNA of the 10kd protein of photosystem II in potato, report on a 100-fold difference in the protein activity level in individual transformed plants. Whereas some plants showed no reduction in protein level, in other transgenic potato plants down-regulation to less than 1% of control activity could be observed. Further transgenic potatoes exhibited all possible degrees of inhibition in between. Such a variable degree of inhibition seems to be common to practically all examples listed in Table I. It should be added that some reports have focused just on those plants which show a high degree of gene suppression.

The variability of the antisense phenotype seems to be linked to a particular antisense locus. Peñarrubia *et al.* (1992) have analyzed the progeny of a transgenic tomato line that contained about four antisense genes directed against the E8 protein. The transgenic antisense plants show an increase in ethylene production, since the E8 protein was found to have a negative effect on ethylene production specifically during ripening of detached fruits. Peñarrubia *et al.* could show that only some but not all the plant progeny that inherited an antisense locus continued to suppress the E8 protein.

The different response to the antisense gene in different transgenic plants obviously must be caused by a "position effect", *i.e.* it depends on the site where the antisense gene had been incorporated into the plant genome. Van der Krol *et al.* (1990a) tried to address this hypothesis but could not correlate a certain phenotype with the insertion of the antisense DNA into a particular chromosome.

The mixed phenotypes amongst different transgenic plants indicates that the border sequences influence the effectiveness of the antisense gene in a quantitative and qualitative way. The potential - and quite speculative - explanations for the observed position effect reflect our limited knowledge on the influence of DNA topology on RNA transcription and post-transcriptional events. Besides a different transcription efficiency due to the sequence context within the chromosome, one could assume different local hnRNP proteins, different composition of hnRNP proteins, different intranuclear transportation pathways of the antisense RNA or folding into variable secondary structures depending on the chromosomal location of the antisense gene.

Table 1

No.	Report	Plant	Target gene (sense gene) name	function involved
1	Ecker and Davis, 1986	carrot protoplasts	chloramphenicl acetyltransferase (*cat*)	no plant gene; model system
2	Rothstein *et al.*, 1988	tobacco	nopaline synthase (*nos*)	T-DNA gene; of *A. tumefaciens*
3	Delauney *et al.*, 1988	tobacco	chloramphenicol acetyltranferase (*cat*)	no plant gene; model system
4	Rodermel *et al.*,	tobacco	ribulose bishosphat carboxylase (RUBISCO) small subunit (*rbc*S)	photosynthesis
5	Sandler *et al.*, 1988	tobacco	nopaline synthase (*nos*)	T-DNA gene. of *A. tumefaciens*
6	Sheehy *et al.*, 1988	tomato	polygalacturonase (*pg*)	fruit ripening
7	Smith *et al.*, 1988	tomato	polygalacturonase (*pg*)	fruit ripening
8	van der Krol *et al.*, 1988a	petunia (tobacco)	chalcone synthase (*chs*)	flavonoid biosynthesis, flower pigmentation
9	Cornelissen and Vandewiele, 1989	tobacco	phosphinitricin acetyl transferase (PAT) (*bar*)	bialaphos resistance
10	Robert *et al.*,, 1988	tobacco	β-glucuronidase (*gus*)	no plant gene; model system
11	Cannon *et al.*, 1990	*Nicotiana plumbaginifolia*	β-glucuronidase (*gus*)	no plant gene; model system
12	Choudhary *et al.*, 1990	alfalfa	chalcone synthase from bean (*chs*)	flavonoid biosynthesis, flower pigmentation
13	Hamilton *et al.*, 1990	tomato	pTOM13 gene (ACC oxidase)	fruit ripening
14	Smith *et al.*, 1990a	tomato	polygalacturonase (*pg*)	fruit ripening
15	Stockhaus *et al.*, 1990	potato	10 kd protein of photosystem II	photosynthesis
16	van der Krol *et al.*, 1990b	petunia	chalcone synthase (*chs*)	flavonoid synthesis
17	Bird *et al.*, 1991	tomato	pTOM5 gene	fruit ripening
18	Oeller *et al.*, 1991	tomato	1-aminocyclopropane-1-carboxylate synthase (ACC synthase)	fruit ripening
19	Visser *et al.*, 1991	potato	granuole-bound starch synthase	reserve starch biosynthesis
20	Knutzon *et al.*, 1992	rape	stearoyl-ACP desaturase	desaturation of stearing acid
21	McGurl *et al.*, 1992	tomato	prosystemin	induction of protease inhibitors
22	Müller-Röber *et al.*, 1992	potato	ADP-glucose pyrophosphorylase	biosynthesis of starch precursor
23	Peñarrubia *et al.*, 1992	potato	E8 gene	fruit ripening
24	Tieman *et al.*, 1992	tomato	pectin methylesterase (PME)	fruit ripening

a. The length of the cDNA which is used as antisense gene is indicated; full: the entire cDNA is expressed as antisense gene; if a cDNA fragment was used it is indicated as "partial", numbers given in parenthesis indicate the part of the cDNA and its length; e.g. : (5'/172) or (5'/65%): the 5' portion, consisting of 172 base pairs or 65% of the DNA is expessed as antisense gene.

b: Only plants with a clearly detectable antisense effect are listed.

c: The activity of the control plant is considered 100%.

d: Steady-state mRNA concentration.

e: Enzyme activity or protein concentration – where metabolites are compared it is indicated by an extra footnote.

m: Maximal inhibition – this level of inhibition is not detected in all plants.

3a: Complete inhibition of enzyme activity was observed in 6 of 32 plants. In 9 other plants inhibition was partial.

3b: No complete inhibition was observed, in 10 out of 18 plants inhibition was partial.

4a: Activity range ot the 5 plants analyzed.

5a: Plants had been analyzed in 9 pools, all pools showed an antisense effect.

6a: Value of one representative plant; run-off transcription showed that the transcription rate of the endogenous PG gene was not affected.

6b: Activity range of several plants.

9a: This represents the activity level of one plant studied in detail. The level of the other 5 plants varied between 3 and 78%.

10a: In each of the 5 plants analyzed in detail, the residual activity was below 10%.

11a: The antisense DNA also included an additional 12 nucleotides of the untranslated leader sequence, resulting in a total length of 41 nucleotides.

promoter	Antisense gene promoter	Antisense gene length[a]	No. of plants trans-formed	detectable antisens effect[b]/ analyzed plants	Activity level in % in antisense versus control plants[c] mRNA[d]	enzyme or protein[e]
CaMV35S, pal, nos	CaMV35S, pal, nos	full	transient assay		n.l.	< 5
nos	CaMV35S	partial (5'/65%)	5	4/5	10-12.5	2-10
		full	32	15/32	n.l.	0[3a]
CaMV19S	CaMV35S	partial (5'/172)	18	10/18	n.l.	n.l.[3b]
endogenous	CaMV35S	full	5	5/5	12-33	37-63[4a]
nos	cab22 (petunia)	partial+full	7-15	9/9[5a]	n.l.	15
endogenous	CaMV35S	full	10	9/10	10[6a]	7-31[6b]
endogenous	CaMV35S	partial (5'/730)	n.l.	3/n/l/	6[illegible]	10[illegible]
endogenous	CaMV35S	full	25	13/25	effect on flower pigmentation class 1: no phenotype (12/25) class 2: reduction or alternation (8/25) class 3: white phenotype (5/25)	
TR'2	CaMV35S	full	6	6/6	25	8[9a]
CaMV35S	CaMV35S	full	10	10/10	n.l.	< 10[10a]
CaMV35S	ca/b *A. thaliana*	partial (5'/29)[11a]	16	13/16	n.l.	0[11b]
endogenous	CaMV35S	partial+full	transient assay		loss of elecitor-induced expression	
endogenous	CaMV35S	partial (5'/n.l.)	n.l.	1/1[13a]	n.l.	13 or 32[13b]
endogenous	CaMV35S	partial (5'/730)[14a]	13	5/5	n.l.	5-51[14b]
endogenous	CaMV35S	full	36	12/36[15a]	1-5[15b]	0.7[illegible]
endogenous	CaMV35S	(5'/704)	20	0/20	class 1: 20	class2: 0; class3[16a]: 0
		(3'/628)	13	4/13	9	3; 1
		(3'/328)	10	0/10	10	0; 0
		(3'/292)	13	1/13	12	1; 0
endogenous	CaMV35S	partial (5'/794)	33	5/5[17a]	n.l.	3-7 or 0.1-2[17b]
endogenous	CaMV35S	full	34	3/34[18a]	n.l.	0.5[18b]
endogenous	CaMV35S	full	46	13/16	n.l.	0-3[19a]
endogenous	napin or ACP carrier protein	full	22	20/22	n.l.	0[illegible]
endogenous	CaMV35S	full	18	11/18	n.l.	< 40, < 30[21a]
endogenous	CaMV35S	partial	n.l.	8/35	n.l.	1.5-17[22a]
endogenous	CaMV35S	full	10	2/10	n.l.	< 1
endogenous	CaMV35S	full[24a]	18	10/10	n.l.	< 10[illegible]

n.l.: not listed in report

11b: In 8 of 13 plants no GUS activity was detectable. In 4 plants it was below 25% and in another plant about 50% compared to the control.
13a: All transformed plants showed an antisense effect. However, the number of plants is not listed – one plant was analysed in further detail.
13b: The activity level refers to the ethylene production rate after wound induction and during fruit ripening, respectively – in homozygous plants the level of ethylene product during fruit ripening was reduced to 3%.
14a: The antisense DNA also contained 50 bases of the untranslated leader sequence.
14b: The activity range of 5 plants. After selfing of the plant which has a 10% activity level, the activity level went down to 1% in the homozygous progeny.
15a: At least 12 plants showed an antisense effect - it is not listed how many of the residual plants showed an antisense effect.
15b: Activity values of 4 plants with the strongest antisense effect.
16a: Same classification as in 9.
17a: 29 out of 33 plants showed a pale phenotype.
17b: The carotenoid and lycopene content is listed, respectively.
18a: 3 plants show marked inhibition.
18b: Ethylene production rate in fruits.
19a: Activity values of 6 plants with the strongest antisense effect.
21a: The activity level or proteinase inhibitor I and II was below 40 and 30% in 11 plants analyzed.
22a: Activity values of 8 plants with the strongest antisense effect.
24a: Genomic DNA was used as antisense DNA. It contained intron sequences and untranslated sequences.

It seems that a duplex of about 40 nucleotides is long enough to fulfill this requirement.

Tieman *et al.* (1992, compare Table I, No 24) could show that also genomic DNA can be used for expression of an antisense RNA. The presence of an intron sequence did not influence the antisense phenotype.

SPECIFICITY OF THE ANTISENSE RNA

Gene suppression by antisense RNA does not require the formation of a perfect duplex RNA. Van der Krol *et al.* (1988c) found that the *chsA* and the *chsJ* genes, which share 86% sequence similarity, are inhibited by the same antisense RNA which was derived from a *chsA* antisense gene. Moreover, inhibition worked as well in the heterologous system when the petunia *chs* antisense gene was expressed in tobacco (van der Krol *et al.*, 1988a). The corresponding tobacco gene shares about 80% sequence similarity (van der Krol *et al.*, 1988c). On the other hand, Peñarrubia *et al.* (1992) analyzed the gene encoding the E8 protein of tomato which shares 51% sequence similarity to the gene encoding the ethylene forming enzyme (EFE). When the corresponding E8 antisense gene had been introduced in tomatoes, no reduction of ethylene production was observed, unlike the situation in tomatoes that expressed the related EFE-directed antisense RNA (Hamilton *et al.*, 1990). In fact, Peñarrubia *et al.* (1992) found that the E8 antisense plants showed an ethylene overproduction. This is because the E8 protein is involved in shutting off ethylene production. Thus, despite the sequence similarity of the two genes, different suppression phenotypes could be observed.

Presumably there is no general rule to what extent two RNAs must be related so that they are suppressed by the same antisense RNA. More important than the degree of sequence similarity is again the length of the possible double-stranded domain.

For the specific suppression of an individual gene, which is for example a part of a multigene family, it might be better therefore, to reduce the size of the antisense RNA and to select a target region that is specific for the target gene.

EXAMPLES OF QUESTIONS THAT CAN BE ADDRESSED BY EXPRESSION OF ANTISENSE RNAs IN PLANTS

Assigning of Gene Function

The antisense technology provides an excellent tool for assigning a biological function to an identified cDNA sequence by creating a mutant phenotype. This was demonstrated by Grierson and co-workers (Hamilton *et al.*, 1990) who identified the "*pTOM13* gene" from a tomato cDNA library that had been prepared from ripening fruit tissue. Based on the correlation of pTOM13 mRNA with ethylene synthesis, the authors assumed that they had identified the ACC oxidase, also called the ethylene forming enzyme (EFE). When the corresponding antisense RNA was expressed in tomato ethylene production was indeed inhibited. Since a secondary effect on ethylene synthesis could not be excluded completely, the corresponding cDNA was expressed in yeast (Hamilton *et al.*, 1991) or in *Xenopus* oocytes (Spanu *et al.*, 1991). In both systems, which normally neither produce 1-amino-cyclopropane-1-carboxylic acid (ACC) nor ethylene, externally applied ACC could be converted to ethylene in the presence of the *pTOM13* gene. These experiments finally confirmed that the correct gene for the EFE enzyme had been isolated. A detailed description of the functional identification of the *pTOM13* gene can be found in the reviews by Bouzayen *et al.* (1992) and by Grierson (1992).

A similar example where a function could be assigned to a cDNA is the "*pTOM5* gene" (Bird *et al.*, 1991). Upon expression of a *pTOM5* antisense RNA, carotenoid biosynthesis was inhibited. It was evident that the *pTOM5* gene is involved in the biosynthesis of carotenoids though it originally could not been determined which individual step of the carotenoid biosynthetic pathway had been blocked. In a further study, Bramley *et al.* (1992) prepared a cell-free system for carotenoid biosynthesis from these transgenic tomato fruit. Using different radioactively labeled carotenoid precursors, they could show that in fruit from the pTOM5 antisense plants the carotenoid biosynthesis pathway was blocked between geranylgeranyl diphosphate (GGDP) and phytoene. However, carotenoid biosynthesis in leaf was not affected.

Also by a functional antisense test, the role of the E8 protein could be elucidated. It was demonstrated that the E8 protein is involved in the negative feedback of ethylene production during fruit ripening and shuts off ethylene production after the respiratory burst (Peñarrubia *et al.*, 1992).

Analysis of Coordinated Gene Expression in Multimeric Complexes

Ribulose-1,5-bisphosphate carboxylase-oxygenase (RUBISCO) is a multimeric protein complex located in the chloroplast consisting of eight small and eight large subunits. The large subunit is encoded by the chloroplast genome, whereas the small subunit is encoded by a nuclear multigene family. Bogorad and co-workers (Rodermel *et al.*, 1988) were interested to study how the stoichiometry of the two subunits is regulated in view of the observation that free subunits do not accumulate and in view of the transcription of their mRNA in different cellular compartments. Suppression of the small subunit by an antisense gene drastically reduced the accumulation of the mRNA and the protein of the small subunit (Rodermel *et al.*, 1988, compare Table I, No 4). The amount of mRNA encoding the large subunit was not affected, however, a corresponding reduction in the protein accumulation of the large subunit was detected. This finding indicated the existence of an uncharacterized sensor system that controls the translation of the large subunit of RUBISCO. Inhibition of the small subunit of RUBISCO also could be used to study its impact on photosynthesis in ambient growth conditions (Quick *et al.*, 1991), on flux-control coefficient for photosynthesis in varying light, CO_2 and air humidity conditions (Stitt *et al.*, 1991) and CO_2 assimilation rates (Hudson *et al.*, 1992).

Analysis of Metabolic Pathways

Müller-Röber *et al.* (1992) intended to analyze whether ADP-glucose is the sole precursor for starch biosynthesis in potato tubers. Using antisense RNA, they suppressed the ADP glucose pyrophospholyase (AGPase) which generates ADP-glucose from glucose-1 phosphate and ATP. In transgenic plants in which the AGPase gene had been effectively suppressed, the tubers contained almost no starch, indicating that starch biosynthesis was blocked completely. Instead of starch, these tubers contained about 30% sucrose and 8% glucose by dry mass.

Identification of Isoenzymes

Using transgenic tomato plants that expressed an antisense RNA against the fruit-specific polygalacturonase (PG), Taylor *et al.* (1990) found that the PG activity during leaf abscission was not inhibited, unlike the PG activity in ripening fruit. This finding indicated that an unrelated PG isoenzyme had been identified. This was substantiated further by immunological techniques and Northern blot analysis.

Applications for Crop Improvement

The application of the antisense technology to crop improvement has just started. There is no doubt that there will be plenty of practical applications for suppressing undesired plant genes. So far, the most spectacular achievements have been made by suppressing several genes that are involved in tomato fruit ripening. Two enzymes involved in ethylene synthesis have been suppressed, the key enzyme of ethylene biosynthesis, the ACC synthase (Oeller *et al.*, 1991 compare Table I, No 18) and the ACC oxidase or ethylene forming enzyme (EFE) (Hamilton *et al.*, 1990, compare Table I, No 14). Inhibition of the ACC synthase prevents the respiratory burst during fruit ripening and mature tomatoes stay green in air. They neither get soft nor do they develop aroma for at least 120 days. Fruit ripening can be induced by treatment with either propylene or ethylene which reverse the antisense phenotype.

Also in tomato, two enzymes that are involved in cell wall degradation have been inhibited: the polygalacturonase (PG) (Sheehy *et al.*, 1988, compare Table I, No 6, 7, 14; Smith *et al.*, 1988 and 1990a) and the pectin methylesterase (PME) (Tieman *et al.*, 1992, compare Table I, No 24). It is anticipated that the inhibition of cell wall degrading enzymes will inhibit or retard fruit softening and therefore improve the shelf-life of the fruits. Such tomatoes already have been tested under field conditions (Kramer *et al.*, 1990). Detailed reviews on fruit ripening and its manipulation by antisense RNA can be found in Schuch *et al.* (1989), Gray *et al.* (1992) and Eckes (1992).

Another example where crops can be "designed", was the seed-specific antisense-mediated inhibition of the stearoyl-ACP desaturase in *Brassica* (Knutzon *et al.*, 1992, compare Table I, No 20). After inhibition of the key enzyme that is responsible for desaturation of fatty acids, the fatty acid composition in rape seed changed dramatically. The seeds contained stearate levels up to 40% which represents a much higher portion of saturated fatty acids. This illustrated the potential of antisense genes for engineering specialized seed oil compositions.

In addition to the above mentioned studies on the inhibition of the chalcone synthase in petunia, which result in altered flower pigmentation, it was shown recently that inhibition of flavonoid biosynthesis can result in male sterility (van der Meer *et al.*, 1992). In that case, the *chs* antisense gene was under the control of the CaMV35S promoter into which an "anther box" had been added. Expression from this tissue-specific promoter caused inhibition of flavonoid biosynthesis in anthers and as a result of that male sterility due to an arrest in male gametophyte development. This study also underlines the fact that the antisense RNA has to be expressed in the corresponding tissue to the target RNA for generating the antisense phenotype.

Is Every Plant Gene a Potential Target for Antisense RNA?

The examples where gene suppression by antisense RNA has been successfully applied (compare Table I) comprise several classes of genes involved in quite different physiological processes, as summarized above. Taking into account that even abundant proteins like RUBISCO can be inhibited, this suggests that in principle, each plant gene is susceptible to the expression of antisense RNA and specifically can be suppressed. However, given the widespread view that "negative" results are not hard data at all, unsuccessful attempts to achieve down-regulation of a plant gene by antisense RNA will, unlike positive examples, not become published in regular reports. The exclusive publication of positive reports potentially might give a false interpretation of the power of the antisense technology in plants. Mol *et al.* (1990) mentioned in a review that attempts so far had failed to down-regulate the petunia chalcone isomerase (CHI) by expression of antisense RNA. This enzyme also is involved in flavonoid biosynthesis, "downstream" of the key enzyme, the chalcone synthase (CHS), which had earlier been shown to be susceptible to antisense RNA (van der Krol *et al.*, 1988a). Fortunately, it seems that the application of antisense RNA leads in most cases to the desired suppression of the target gene. However, in view of the greatly variable phenotypes in plants that express antisense RNA, a large number of transgenic plants might have to be tested in order to identify the desired mutant phenotype. An indication of the extent to which the antisense technology is applicable in plants could be deduced from the so far unpublished work of Willmitzer and co-workers (personal communication). They have targeted in total 12 potato genes by antisense RNA, leading in 11 cases to the desired suppression. Currently, it is too early to conclude finally whether that this is a representative proportion.

The Phenomenon of Co-Suppression.

It was quite a surprise when it was found that the expression of a transgene in sense orientation can result also in the suppression of the corresponding endogenous gene, similar to the suppression that is achieved by expression of antisense RNA (Napoli *et al.*, 1990; Smith *et al.*, 1990b; van der Krol *et al.*, 1990c). This phenomenon has been termed co-suppression and the potential reasons for its occurrence have been discussed in detail in the reports mentioned and in a review by Jorgensen (1990). Gene suppression by a transgene that expresses the sense RNA could indicate that the antisense approach also might involve other mechanisms of inhibition than duplex RNA formation. However, Grierson *et al.* (1991) discussed the idea that co-suppression actually might be caused by transcription of the antisense strand from the integrated sense gene. For a more extensive discussion of co-suppression, see also Mol *et al.* (1991) and Jorgensen (1991).

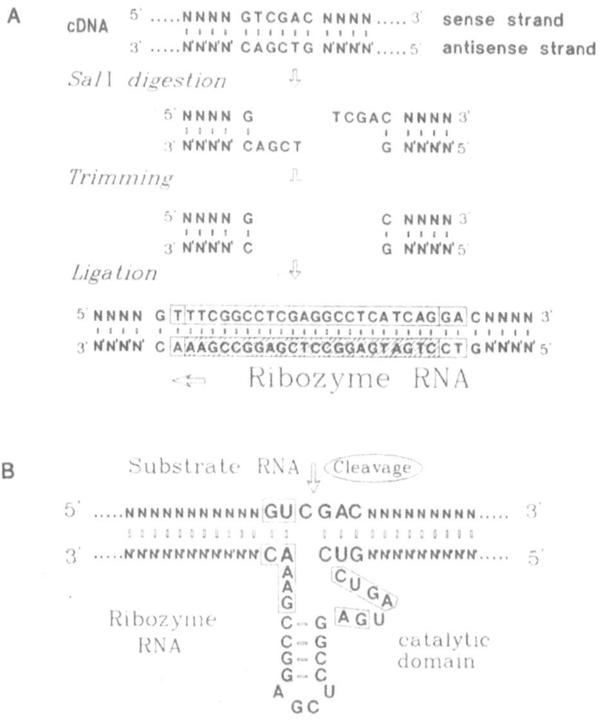

Figure 2. Strategy for converting a cDNA that contains a *Sal*I recognition sequence into a ribozyme encoding DNA, according to Tabler and Tsagris (1991). Part **A** shows sense and antisense strand of the cDNA with its *Sal*I recognition sequence G/TCGAC detailed. The sense strand corresponds to the RNA that will be targeted by ribozyme attack. After cleavage with *Sal*I and trimming the protruding ends, a DNA cassette is inserted which comprises the catalytic domain of a hammerhead ribozyme (hatched part) plus additional terminal nucleotides that replace three of the four substrate-specific nucleotides, previously removed by trimming. From the resulting DNA construct a ribozyme RNA can be synthesized if the antisense strand is expressed. Part **B** shows the interaction of the resulting ribozyme RNA transcript with the substrate RNA in the hammerhead conformation. The boxed sequences are assumed to participate in the cleavage reaction. The site of cleavage is indicated. In this case the catalytic domain had been modified to contain a *Stu*I (AGG/CCT) site and an *Xho*I site (C/TCGAG). The insertion technique can be improved if DNA cassettes are used that contain a selectable marker gene (compare Tabler and Tsagris, 1991).

binding. The specificity of binding does not increase necessarily with the length of the flanking sequences (compare Herschlag, 1991).

Unlike "ordinary" antisense RNAs, ribozyme RNAs not only bind to their target but cause irreversible cleavage. In a truly catalytic reaction, one ribozyme molecule would cleave more than one substrate molecule. Such a mode of reaction would require three phases: (i) binding of the ribozyme to the target, (ii) cleavage of the target RNA and (iii) dissociation of the ribozyme RNA. In order to facilitate dissociation, the flanking sequences must not be too long. A number of examples are described where such an approach has been used for gene suppression (Cameron and Jennings, 1989; Cotten and Birnstiel, 1989; Sarver *et al.*, 1990; Saxena and Ackerman, 1990; Scanlon *et al.*, 1991; Sioud and Drlica, 1991). Recently, a hammerhead ribozyme which was directed against the mRNA of the neomycin phosphotransferase gene (npt) had been tested in plant cells (Steinecke *et al.*, 1992). For this purpose, two DNAs encoding the ribozyme and the target RNA respectively, were cotransfected in tobacco protoplasts. Although it is usually hard to detect, the authors could visualize the cleaved substrate RNA, indicating that the ribozyme approach indeed worked *in vivo*.

A potential problem of expressing ribozymes is that the complementary sequences have to be relatively small if catalytic turn-over is desired. However, sequences that are too short may not bind efficiently to the target RNA since they cannot compete with its secondary structure. Furthermore, small molecules might not be sufficiently stable *in vivo*. For this reason, we started recently to combine the antisense and ribozyme technologies by testing modified antisense RNAs that contain in addition one or several catalytic hammerhead domains. Unlike the "small" ribozyme RNAs such "catalytic antisense RNAs" will not be able to dissociate after cleavage of the target RNA and inactivate the substrate in a stoichiometric rather than a catalytic manner. Unlike "conventional" antisense RNAs, however, the catalytic antisense RNAs are able to cleave their bound target RNA, resulting in irreversible inactivation. For convenient generation of catalytic antisense RNAs, we have developed a technique that allows the insertion of a ribozyme cassette into certain restriction sites of a cDNA (Tabler and Tsagris, 1991). The resulting recombinant DNA construct allows the synthesis of antisense RNA carrying the catalytic domain of the hammerhead ribozymes.

About 25 different restriction recognition sequences can be used, which overlap with a cleavable motif for the hammerhead ribozymes. For example, each *Sal* I-site (G/TCGAC) contains the GUC cleavage motif against which a ribozyme can be directed (Fig. 2). After *Sal* I digestion of the cDNA and removal of the protruding ends, a *Sal* I-specific ribozyme DNA cassette is inserted, containing the catalytic ribozyme domain plus three of the four nucleotides that had been removed by trimming. Transcription delivers the desired antisense-ribozyme. For further details on "selectable ribozyme cassettes" and on the restriction sites that can be used for insertion of such cassettes see Tabler and Tsagris (1991).

So far, we have not tested this approach of modified, catalytic antisense RNAs for gene suppression in plants. However, in collaboration with the laboratory of G.Sczakiel, catalytic domains were incorporated into an HIV-1 antisense RNA. In transfection experiments it was shown that the catalytic antisense RNA was about four- to seven-fold more effective in inhibiting the HIV-1 replication than the corresponding antisense control (Sczakiel and Tabler, unpublished results). Furthermore, a catalytic RNA which had lost its cleavage activity due to a one nucleotide deletion, only had an inhibitory effect comparable to the antisense RNA. This indicates that the cleavage activity of the catalytic antisense RNAs could improve the antisense effect.

It will have to be verified whether the modification of antisense RNA by incorporation of a catalytic ribozyme domain also results in improvement of the antisense technology in plants.

ACKNOWLEDGEMENT

The author is grateful to Drs. Don Grierson and Andrew Hamilton for critical reading the manuscript.

REFERENCES

Bird CR, Ray JA, Fletcher JD, Boniwell JM, Bird AS, Teulieres C, Blain I, Bramley PM, Schuch W (1991) Using antisense RNA to study gene function: inhibition of carotenoid biosynthesis in transgenic tomatoes. *Biotechnology* 9:635-639.

Blomberg P, Wagner EG, Nordström K (1990) Control of replication of plasmid R1: the duplex between the antisense RNA, CopA, and its target, CopT, is processed specifically *in vivo* and *in vitro* by RNase III. *EMBO J* 9:2331-2340.

Bourque JE, Folk WR (1992) Suppression of gene expression in plant cells utilizing antisense sequences transcribed by RNA polymerase III. *Plant Mol Biol* 19:641-647.

Bouzayen M, Hamilton A, Picton S, Barton S, Grierson D (1992) Identification of genes for the ethylene-forming enzyme and inhibition of ethylene synthesis in transgenic plants using antisense genes. *Biochemical Society Transactions* 20:76-79.

Bramley P, Teulieres C, Blain I, Bird C, Schuch W (1992) Biochemical characterization of transgenic tomato plants in which carotenoid synthesis has been inhibited through the expression of antisense RNA to pTOM5. *Plant J* 2:343-349.

Bruening G (1989) Compilation of self-cleaving sequences from plant virus satellite RNAs and other sources. *Methods Enzymol* 180:546-558.

Bruening G, Buzayan JM, Hampel A, Gerlach WL (1988) Replication of small satellite RNAs and viroids: possible participation of non-enzymatic reactions. *In*: Viroids and RNA Recombination, RNA Genetics, Domingo E, Holland JJ, Alquist P (eds), vol II, Retroviruses, , CRC Press, Boca Raton, USA, pp 127-145.

Cameron FH, Jennings PA (1989) Specific gene suppression by engineered ribozymes in monkey cells. *Proc Natl Acad Sci USA* 86:9139-9143.

Cannon M, Platz J, O'Leary M, Sookdeo C, Cannon F (1990) Organ-specific modulation of gene expression in transgenic plants using antisense RNA. *Plant Mol Biol* 15:39-47.

Cech TR (1987) The chemistry of self-cleaving RNA and RNA enzymes. *Science* 231:1532-1539.

Cesareni G, Helmer-Citterich M, Castagnoli L (1991) Control of ColE1 plasmid replication by antisense RNA.

Chapman S, Kavanagh T, Baulcombe D (1992) Potato virus X as a vector for gene expression in plants. *Plant J* 2:549-557.

Choudhary AD, Kessmann H, Lamb CJ, Dixon RA (1990) Stress response in alfalfa (*Medicago sativa* L.) IV. Expression of defense gene constructs in electroporated suspension cell protoplasts. *Plant Cell Rep* 9:42-46.

Cohen JS (1989) Oligodeoxynucleotides: antisense inhibitors of gene expression, Macmillan/CRC Press, London/Miami.

Cohen JS (1992) Chemically modified oligodeoxynucleotide analogs as regulators of viral and cellular gene expression. *In*: Gene Regulation: Biology of Antisense RNA and DNA, Erickson RP, Izant JG (eds), Raven Press Ltd, New York, pp 247-259.

Cornelissen M, Vandewiele M (1989) Both RNA level and translation efficiency are reduced by antisense RNA in transgenic tobacco. *Nucleic Acids Res* 17:833-843.

Cotten M Birnstiel ML (1989) Ribozyme mediated destruction of RNA *in vivo*. *EMBO J* 8:3861-3866.

Cuozzo M, O Connell KM, Kaniewski W, Fang R-X, Chua N-H, Tumer NE (1988) Viral protection in transgenic tobacco plants expressing the cucumber mosaic virus coat protein or its antisense RNA. *Biotechnology* 6:549-557.

Day AG, Bejarano ER, Buck KW, Burrell M, Lichtenstein CP (1991) Expression of an antisense viral gene in transgenic tobacco confers resistance to the DNA virus tomato golden mosaic virus. *Proc Natl Acad Sci USA* 88:6721-6725.

Delauney AJ, Tabaeizadeh Z, Verma DP (1988) A stable bifunctional antisense transcript inhibiting gene expression in transgenic plants. *Proc Natl Acad Sci USA* 85:4300-4304.

Ecker JR, Davis RW (1986) Inhibition of gene expression in plant cells by expression of antisense RNA. *Proc Natl Acad Sci USA* 83:5372-5376.

Eckes P (1992) Inhibition of fruit ripening by antisense-RNA technology. *Angew Chem Int Ed Engl* 31:175-177.

Edgington SM (1992) Ribozymes: stop-making sense. *Biotechnology* 10:256-262.

Eguchi Y, Taleo I, Tomizuwa J (1991) Antisense RNA. *Annu Rev Biochem* 60:631-652.

Forster AC, Symons RH (1987a) Self-cleavage of plus and minus RNAs of a virusoid and a structural model for the active sites. *Cell* 49:211-220.

Forster AC, Symons RH (1987b) Self-cleavage of virusoid RNA is performed by the proposed 55-nucleotide active site. *Cell* 50:9-16.

Gray J, Picton S, Shabbeer J, Schuch W, Grierson D (1992) Molecular biology of fruit ripening and its manipulation with antisense genes. *Plant Mol Biol* 19:69-87.

Green PJ, Pines O, Inouye M (1986) The role of antisense RNA in gene regulation. *Annu Rev Biochem* 55:569-597.

Grierson D (1992) Control of ethylene synthesis and ripening by sense and antisense genes in transgenic plants. *Proc Roy Soc Edinburgh* 99B:79-88.

Grierson D, Fray RG, Hamilton AJ, Smith CJ, Watson CF (1991) Does co-suppression of sense genes in transgenic plants involve antisense RNA? *Trends Biotechnol* 9:122-123.

Hamilton AJ, Lycett GM, Grierson D (1990) Antisense gene that inhibits synthesis of the hormone ethylene in transgenic plants. *Nature* 346:284-287.

Hamilton AJ, Bouzayen M, Grierson D (1991) Identification of a tomato gene for the ethylene-forming enzyme by expression in yeast. *Proc Natl Acad Sci USA* 88:7434-7437.

Haseloff J, Gerlach WL (1988) Simple RNA enzymes with new and highly specific endoribonuclease activities. *Nature* 334:585-591.

Hélène C (1992) Control of gene expression by antisense and antigen oligonucleotide-intercalator conjugates. *In*: Gene Regulation: Biology of Antisense RNA and DNA, Erickson RP, Izant JG (eds), Raven Press Ltd, New York, pp 109-118.

Hélène C, Toulmé J-J (1990) Specific regulation of gene expression by antisense, sense and antigene nucleic acids. *Biochim Biophys Acta* 1049:99-125.

Hemenway C, Fang R-X, Kaniewski WK, Chua N-H, Tumer NE (1988) Analysis of the mechanism of protection in transgenic plants expressing the potato virus X coat protein or its antisense RNA. *EMBO J* 7:1273-1280.

Herschlag D (1991) Implications of ribozyme kinetics for targeting the cleavage of specific RNA molecules *in vivo*: more isn't always better. *Proc Natl Acad Sci USA* 88:6921-6925.

Hobbs SL, Kpodar P, DeLong CM (1990) The effect of T-DNA copy number, position and methylation on reporter gene expression in tobacco transformants. *Plant Mol Biol* 15:851-864.

Hudson GS, Evans JR, von Caemmerer S, Arvidsson YB, Andrews TJ (1992) Reduction of ribulose-1,5-bisphosphate carboxylase/oxygenase content by antisense RNA reduces photosynthesis in transgenic tobacco plants. *Plant Physiol* 98:294-302.

Izant JG, Weintraub H (1984) Inhibition of thymidine kinase gene expression by antisense RNA: a molecular approach to genetic analysis. *Cell* 36:1007-1015.

bisphosphate carboxylase-oxygenase in transgenic tobacco transformed with "antisense" rbcS. II. Flux-control coefficients for photosynthesis in varying light, CO_2, and air humidity. *Planta* 183:555-556.

Stockhaus J, Höfer M, Renger G, Westhoff P, Wydrzynski T, Willmitzer L (1990) Anti-sense RNA efficiently inhibits formation of the 10kd polypeptide of photosystem II in transgenic potato plants: analysis of the role of the 10 kd protein. *EMBO J* 9:3013-3021.

Symons RH (1989) Self-cleavage of RNA in the replication of small pathogens of plants and animals. *Trends Biochem Sci* 14:445-450.

Symons RH (1990) The fascination of low molecular weight pathogenic RNAs. *Sem Virology* 1:75-81.

Tabler M, Tsagris M (1990) Viroid replication mechanisms. *In*: Recognition and Response in Plant-virus Interactions, Fraser RS (ed), *NATO ASI Series*, vol. H41:185-205.

Tabler M, Tsagris M (1991) Catalytic antisense RNAs produced by incorporating ribozyme cassettes into cDNA. *Gene* 108:175-183.

Taylor JE, Tucker GA, Lasslett Y, Smith CJ, Arnold CM, Watson CF, Schuch W, Grierson D, Roberts JA (1990) Polygalacturonase expression during leaf abscission of normal and transgenic tomato plants. *Planta* 183:133-138.

Tieman DM, Harriman RW, Ramamohan G, Handa AK (1992) An antisense pectin methylesterase gene alters pectin chemistry and soluble solids in tomato fruit. *Plant Cell* 4:667-679.

Uhlmann E, Peyman A (1990) Antisense oligonucleotides: a new therapeutic principle. *Chem Rev* 90:543-584.

Van der Krol AR, Lenting PE, Veenstra J, van der Meer IM, Koes RE, Gerats AG, Mol JN, Stuitje AR (1988a) An anti-sense chalcone synthase gene in transgenic plants inhibits flower pigmentation. *Nature* 333:866-869.

Van der Krol AR, Mol JN, Stuitje AR (1988b) Modulation of eukaryotic gene expression by complementary RNA or DNA sequences. *BioTechniques* 6:958-976.

Van der Krol AR, Mol JN, Stuitje AR (1988c) Antisense genes in plants: an overview. *Gene* 72:45-50.

Van der Krol AR, Mur LA, de Lange P, Gerats AG, Mol JN, Stuitje AR (1990a) Antisense chalcone synthase genes in petunia: visualization of variable transgene expression. *Mol Gen Genet* 220:204-212.

Van der Krol AR, Mur LA, de Lange P, Mol JN, Stuitje AR (1990b) Inhibition of flower pigmentation by antisense CHS genes: promoter and minimal sequence requirements for the antisense effect. *Plant Mol Biol* 14:457-466.

Van der Krol AR, Mur LA, Beld M, Mol JN, Stuitje AR (1990c) Flavonoid genes in petunia: addition of a limited number of gene copies may lead to a suppression of gene expression. *Plant Cell* 2:291-299.

Van der Meer IM, Stam ME, van Tunen AJ, Mol JN, Stuitje AR (1992) Antisense inhibition of flavonoid biosynthesis in petunia anthers results in male sterility. *Plant Cell* 4:253-262.

Visser RG, Somhorst I, Kuipers GJ, Ruys NJ, Feenstra WJ, Jacobsen E (1991) Inhibition of the expression of the gene for granule-bound starch synthase in potato by antisense constructs. *Mol Gen Genet* 225:289-296.

Wickstrom E (1992) Antisense DNA therapeutics neutral analogs and their stereochemistry. *In*: Gene Regulation: Biology of Antisense RNA and DNA, Erickson RP, Izant JG (eds), Raven Press Ltd, New York, pp 119-132.

PHYTOACTIVE POLYMERS: NEW SYNTHETIC PLANT GROWTH REGULATORS

Aristides M. Tsatsakis[1] and **Michael I. Shtilman**[2]

[1] Department of Toxicology
University of Crete,
P.O.Box 1393, Heraklion, GREECE
[2] Department of Polymer Science
Mendelejev Institute
Moscow 125190, RUSSIA

INTRODUCTION

Endogenous plant substances play an important role in plant life. Phytohormones (including the most extensively studied auxins, gibberellins, cytokinins, ethylene and its precursor 1-aminocyclopropane-1-carboxylic acid) as well as other substances, (particularly fusicoccin, abscisic acid, brassinosteroids, aliphatic polyamines and plant cell wall oligosaccharides), are involved in the endogenous regulation of plant growth (Chailahyan, 1982; Kogl, 1934; Nickell, 1982; Tran Thanh Van *et al.*, 1985). All of the above biologically active substances are low molecular weight substances that belong to different classes of organic compounds, namely carboxylic acids (e.g. auxins, gibberellins, abscisic acid), 6-N-substituted adenins (cytokinins), steroids (brassinosteroids), etc.

The discovery, in the 1920's, of the biological activity of 3-indoleacetic acid, heteroauxin, obtained from plant extract, gave impetus to multiple investigations concerning exogenous regulation of plant growth (Kogl, 1934). Studies in this direction have led to the theoretical principles for the application of such compounds, referred to as plant growth regulators (PGRs), i.e. chemical substances with the ability to regulate plant growth. These regulators include phytohormones and other natural substances (e.g. vitamins, nucleotides) as well as many of their synthetic analogues and other synthetic compounds for plant growth. Over time plant growth regulators have gained some practical relevance (Kidd *et al.*, 1988; Nickell, 1982).

More widespread use of these PGRs has been limited by a number of factors. For example, the preparation of a number of these substances requires complex technology, in-

Morphogenesis in Plants, Edited by K.A. Roubelakis-Angelakis
and K. Tran Thanh Van, Plenum Press, New York, 1993

cluding, in some cases, microbiological methods. Once prepared, some of these substances are difficult to apply due to peculiarities in their biological and physical-chemical properties. As a rule, the most active compounds in terms of their biological effects are characterized by a narrow region of stimulating concentrations and doses. If these concentrations or applied doses are exceeded, then plants rather than being stimulated may be inhibited even to the point of death (herbicidic effect). In addition, many PGRs, especially those of natural origin or their synthetic analogues, are biodegradable, and very high doses are required to achieve the desired effect. Higher application doses are needed as well when preparations are washed away easily or are volatile, and the hydrophobic nature of some PGRs is a limiting factor in their application. Finally, many PGRs are quite toxic to humans and other organisms.

On the basis of the above, it can be concluded that the development of modified PGRs which would not have such practical disadvantages, or in which such disadvantages would be minimized, is of great importance to agronomy.

SYSTEMS WITH CONTROLLED RELEASE OF BIOACTIVE MATERIAL

For the last twenty years, increasing attention has been focused on systems with controlled release of bioactive compounds to solve many applied problems in medicine, biotechnology, agronomy, and other fields. Such systems are based on various preparation forms and are able to release a bioactive compound at a certain rate. The rate obtained is dependent mainly on the general construction of the system and in particular, the structure of its ingredients.

The theory of controlled release and its applications has been researched extensively. A number of monographs and reviews have been published (Kydonieus, 1980; Leninger and Pepas, 1983), large international conferences have been conducted (Kellaway, 1991; Kopecek, 1992; Lewis, 1981) an active international society (Controlled Release Society, Inc) and a specialized cited journal (J Controlled Release) exists.

A great number of different systems for controlled release have been investigated, primarily based on two mechanisms of release, i.e. diffusion and cleavage of chemical bonds. Controlled release systems may be prepared in different forms. The active ingredients may be modified with various polymeric materials to give vesicles, micro and macrocapsules, nanoparticles, sheets, gels, dusts, fibers, filaments, membrane coating systems, and systems that are water soluble or soluble in organic solvents.

Some examples of controlled release systems for agriculture are known in the scientific literature. Many of these systems concern the modification of fertilizers and pesticides and are based on the diffusion mechanism of bioactive substance release (Allan *et al.*, 1971; Kydonieus, 1980; Wilkins, 1982). Investigations concerning the creation of agricultural controlled release systems based on hydrolysis of chemical bonds between bioactive compound and polymer carrier have been undertaken. In most of these cases, polymeric herbicides and the biocides with controlled release action were prepared (Allan *et al.*, 1973, 1980; Korshak and Shtilman, 1984; McCormick *et al.*, 1983; Shtilman, 1985). The above principles for bioactive compound modification attracted attention to the construction of polymeric derivatives of plant growth regulators (Allan *et al.*, 1979; Miller *et al.*, 1985; Shtilman, 1983, 1992; Tsatsakis *et al.*, 1990) referred to as phytoactive polymers (Shtilman, 1985, 1987; Tsatsakis *et al.*, 1989, 1990, 1991).

PHYTOACTIVE POLYMERS. PRINCIPLES OF CONSTRUCTION

In general, the macromolecule of a phytoactive polymer is a high molecular weight system in which the residue of bioactive compound (i.e. PGR) is attached to the polymeric chain by a hydrolysable chemical bond. The plant growth regulator residues may be introduced either indirectly onto the chain, attached to the terminal linkages of the polymeric chain, or attached to the side groups of the main polymeric chain (Allan *et al.*, 1979; Korshak and Shtilman, 1984; Miller and Yahiaoui, 1985; Shtilman, 1983; Tsatsakis *et al.*, 1990).

This third way of linking the regulator s residue to a polymeric chain (Fig. 1) has been shown to be the most promising. In this case both the content of bioactive compound in the polymer and the properties of the overall polymeric system may be regulated during synthesis. The regulator residues may be attached either to the side polymer linkage or through a spacer.

The hydrolysable bond between polymeric chain and PGR residues may be ionic, covalent (Fig. 1) or of coordination character. The functional moieties most often used for covalent bonding are esters (Fig. 1), amides, urea, carbamates etc. The type of hydrolysable bond depends upon the reactive groups that the particular low molecular weight bioactive compound structure contains.

In addition to linkages with the low molecular weight regulator residues, the phytoactive polymers may contain other linkages in order to provide the polymeric system with the appropriate hydrophilicity (e.g. containing hydroxylic groups, amidic groups, pyrrolli-

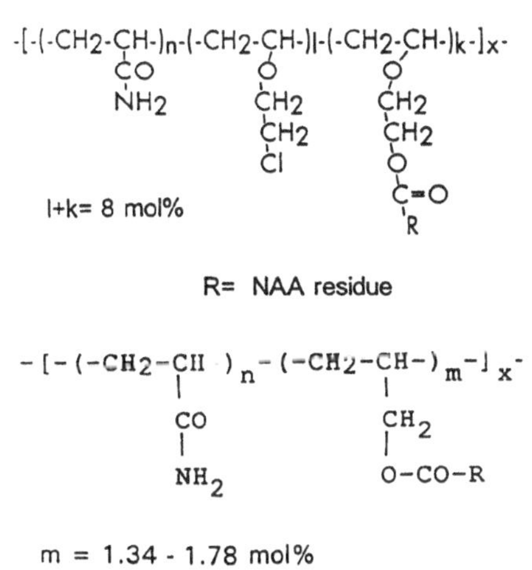

Figure 1. The structures of two phytoactive polymers containing 1-naphthylacetic acid (NAA) residues (Tsatsakis *et al.*, 1990). Upper: copolymer carrier of acrylamide and vinylchloroethyl ether; lower: copolymer carrier of acrylamide and allylic ester.

were used as the low molecular weight bioactive compound. Some of these acids, in addition to regulating plant growth activity, may display biocidic and herbicidic activity (Allan *et al.*, 1971). Early studies on the synthesis of polymeric derivatives of phytohormonal auxins, such as 3-indolylacetic acid, heteroauxin, their synthetic analogues 3-indolylpropionic acid, 3-indolylbutyric acid and 1-naphthylacetic acid were, however, rather poor.

When abilities for the preparation of polymeric systems containing residues of auxins were examined, it was found that the unsaturated esters of aryloxyalkanecarboxylic acid show good ability for homo- and copolymerization (Rehab *et al.*, 1991). The unsaturated derivatives of the arylalkanecarboxylic acids are characterized by a lower ability to homopolymerization (Korshak *et al.*, 1980). Despite that, two examples of copolymerization reactions involving unsaturated derivatives of 1-naphthylacetic and 3-indolylacetic acids as co-monomers are cited (McCormick *et al.*, 1986, 1988).

A considerable number of investigations were devoted to the coupling of the auxinic class carboxylic acids or their derivatives (chloroanhydrides or salts) with natural or synthetic polymeric carriers containing various reactive groups. In cases where the preparation of the polymeric esters was based on polymer carriers that contained hydroxy groups, the chloroanhydrides of auxins usually were involved in the reaction. Natural polymers containing hydroxy groups (e.g. cellulose, starch, alginic acid, dextran, chitin, chitosan, and also lignin and various wood waste) for immobilization of auxins containing indole cycle have not been studied. Hence, the use of the above natural compounds as carriers was documented for the immobilization of other haloidaryl-oxycarboxylic acids (Allan *et al.*, 1971, 1973) and 1-naphthylacetic acid (Allan, 1967).

The reactions of 1-naphthylacetic and 3-indolylacetic acid chloroanhydrides with synthetic polymers containing hydroxy groups, such as, polyvinylalcohol and poly(N-hydroxymethyl-maleimide) (Korshak *et al.*, 1977, 1980), and also of 2,4-dichlorophenoxyacetic and 2-methyl-4-chlorophenoxyacetic acid chloroanhydrides with gels of 2-hydroxyethyl-methacrylate, acrylamide and methylene-bis-acrylamide copolymers were also described (Issa *et al.*, 1990).

The nonactivated carboxylic acids from the class of auxins, in reaction with polymers containing hydroxy groups rarely were involved. One example, however, was the esterification of the poly-vinylalcohol with 2,4-dichlorophenoxyacetic acid, which was achieved in melting conditions. When polymeric carriers containing activated haloidalkyl groups (e.g. chloromethylated polystyrene, poly-N-chloromethylmaleimide, polyvinylchloromethylketone or copolymers of vinyl-2-chloroethyl esters) were used, then the low molecular bioactive compounds were involved in the synthesis in nonactivated form, such as salts of acids or as the acids themselves. In this case, the reaction was carried out in the presence of hydrogen chloride acceptors (Jacubke *et al.*, 1973; Korshak *et al.*, 1980).

Polymeric derivatives of a broad number of auxins indole containing, haloidaroxyalkanecarboxylic, naphthylacetic and phenylacetic acids have been prepared by their reaction with polymeric epoxides (for example polyglycidylmethacrylate). In this instance the acids have been introduced in the reaction directly as compounds containing carboxylic groups (Korshak *et al.*, 1978, 1979, 1982, 1985, 1988; Shtilman *et al.*, 1977).

The polymeric esters of the auxins which have been (in nonactivated forms) prepared in the above cases displayed a different ability to release the biologically active compound

(Shtilman and Tsatsakis, 1990). Other polymeric carriers and other reactions, which resulted in the formation of ionic or coordination bonds between the polymer-carrier and the bioactive ligand, have not been often used in the synthesis of the polymeric auxins (Allan *et al.*, 1980; McCormick *et al.*, 1983; Shtilman, 1985).

Polymeric Derivatives of Gibberellic Acid

The most widely researched representative of the PGRs class of gibberellins is gibberellic acid (Gibberellin A3). Some processes for the preparation of polymeric derivatives of gibberellic acid already have been described. In connection with these examples the ability to use the bonded to agarose gibberellic acid as immobilized ligand for affinity sorbents should be mentioned (Kramel, 1975). At the same time, polymeric derivatives of this acid suitable for use as PGRs were prepared by reaction with it or its potassium salt with a polymer carrier containing epoxide or chloroalkyl groups. In these cases, products both soluble and insoluble in water were formed (Muromtsev *et al.*, 1990). By reaction of the gibberellic acid with poly-2-methyl-5-vinyl-1-pyridinemethylenepentamethyldisiloxane, polymeric salts showing low biological activity were prepared (Kolesnik *et al.*, 1990).

Polymers with Cytokininic Activity

Polymeric derivatives of cytokinins which induce in particular the process of cell division are described in a series of scientific papers. The main representatives of these polymers are derivatives of adenine, one of the nucleic bases. The immobilization of a series of 6-N-substituted adenins on polysaccharides (cellulose and starch) has been achieved by previous treatment of the polysaccharides with phosgene in order to introduce in them the reactive chlorformiate groups (Bitner *et al.*, 1977, 1979).

Polymeric derivatives of cytokinins also have been obtained by attaching 6-N-substituted adenins to polymeric carriers containing isothiocyanate groups. In the above, the regulator residues were linked to a polymeric chain by a thiourea spacer. Among other polymers with cytokininic activity, one should mention the polymeric derivatives of 6-[N-(2,3-dihydroxypropyl)amino]purine, of 5'-adenosinephosphates and N^9-(2,3-dihydroxypropyl) aminopurine. The first of the above polymers was prepared by reaction of the low molecular weight cytokinin with a polymer containing acetale groups, while the second and the third polymer were prepared by reaction of 5'-adenosinephosphates and adenine with polyglycidylmethacrylate (Korshak *et al.*, 1982, 1983; Shtilman *et al.*, 1981). The substitution reaction of the multifunctional molecule of adenosinephosphates when treated with the oxiranic cycle (of polyglycidylmethacrylate) proceeds mainly through the hydroxy group of the adenosine phosphonic acid residue. The above was confirmed by analysis of the products of a model reaction between adenosin-5'-phosphate and glycidol.

Polymeric Derivatives of Ethylene's Producents

Phytohormonal ethylene plays a very important role in plant life. The possibility of its immobilization in polymeric systems containing carbonyls of metals bonded to them in order to obtain growth regulating systems is indicated in the literature (Allan *et al.*, 1992). The ethylene's precursor in plant organisms, 1-aminocyclopropane-1-carboxylic acid (ACC), during exogenous application in plants displayed biological effect similar to that of ethylene (Shtilman *et al.*, 1992). The synthesis of the series of polymeric derivatives of ACC possessing biological activity was already synthesized. These high molecular weight compounds have been prepared by various synthetic methods, such as, the copolymerization of N-methacryloylate derivative of the acid with N-vinylpyrrolidone, the reaction of 1-isocyanatocyclopropyl carboxylic acid methyl ester with polyvinylalcohol and the reaction of 1-aminocyclopropane-1-carboxylic acid with copolymers of maleic anhydride (Kornakov *et al.*, 1992). A polymeric salt of ethylene's producent 2-chloroethyl-phosphonic acid was obtained by reaction of the last with polymer containing basic groups.

Polymeric Derivatives of Other Regulators

Some attempts to prepare polyvinyl esters of electric and sorbinic acids have been described (Korshak *et al.*, 1976, 1977). The above acids in their low molecular weight form possess certain plant growth regulating activity. The polymeric esters of the vitamin 3-pyridinecarbonic (nicotinic) acid also have been found to have notably high biological activity. These polymeric PGRs have been prepared by the acylation reaction of polyvinylalcohol with the chloroanhydride of the acid (Korshak *et al.*, 1982, 1981) or by the reaction of the acid itself with polyglycidylmethacrylate. Different authors have referred to the preparation of biologically active polymeric derivatives of substituted acetylenes, copolymers of bis-trialkylstannyl maleinates and polymer based on N,N'- dimethylhydrazide of the electric acid (Georgieva *et al.*, 1977). The preparation of the polymeric salt of polyacrylate acid and the N-oxide alkaloid of lutidine has been achieved by photoinitiated polymerization of the corresponding monomers.

The biological activity of some synthesized polymeric forms of fusicoccin derivatives has also been mentioned. In this section it should be mentioned that no references to the synthesis of polymeric derivatives of the abscisic acid or its derivatives that display biological activity have been found in the scientific literature. Despite that, examples of abscisic acid immobilization on polymers in order to obtain affinic sorbents are known (Lehmann *et al.*, 1979).

PROPERTIES OF THE PHYTOACTIVE POLYMERS

Solubility

Subject to their chemical structure, the phytoactive polymers are soluble or insoluble in water or organic solvents powdery preparations. The modification of the low molecular weight biologically active compound to its polymeric form contributes to the alteration capabilities of its solubility level. By increasing the number of the lyophilic groups, which are introduced in the polymeric chain and also the level of their lyophilicity, water soluble forms of the PGRs, even for such hydrophobic compounds as the fusicoccin, may be obtained. The water solubility of the final polymeric products with plant growth regulating activity and controlled release of the bioactive low molecular weight compound is very important in terms of application abilities, particularly when traditional agronomy technologies are used. In addition, the lack of solubility of phytoactive polymers makes their uniform distribution on the treated objects impossible especially as very low doses of the biologically active polymers are usually used.

In relation to the above facts, the application of controlled release PGR preparations, in cases where the release of the bioactive compound was based on diffusion mechanisms evoked certain complexities. Such preparations containing adenins, gibberellic acid, 2,4-dichlorophenoxyacetic acid, benzoic acid, 1-naphthylacetic acid have been reported. In these preparation forms, the active compounds have been included inside microporous cellulose fibers or dissolved in polymeric matrices (Allan *et al.*, 1992).

By way of contrast in some particular cases when the low molecular weight bioactive compound is washed away easily, its incorporation into the polymeric carrier may be necessary in order to decrease the water solubility of the final product.

Hydrolysis Ability

A very important characteristic of the phytoactive polymers is their ability gradually to release the low molecular weight bioactive compound by hydrolytic breaking down of the bond between the bioactive substance residues which are contained in the polymer and the polymeric chain. There are few studies in the literature in which the plant regulating activity of polymers that do not contain immobilized low molecular weight regulators have been examined. Nevertheless it is still not clear in these cases if the appearance of the biological effect is due to the polymers themselves, to the presence of little amounts of low molecular weight admixtures within them or to any other causes.

It is important to note the relative role of chemical and enzymatic hydrolysis in the phytoactive polymer function still has not been researched well. Also, investigations concerning the penetration of the released bioactive low molecular weight compound in the plant and the peculiarities of its involving the processes of plant metabolism have not been conducted. Hence, a noticeable number of scientific investigation works have been devoted to hydrolysis studies of the polymeric PGR derivatives in model conditions. From the data already obtained from these studies, conclusions may be drawn about the substantial influence of the final polymeric system chemical structure to the release ability of the low molecular weight bioactive compound from the polymer.

It may be stated that the main factors which influence the release rate of the low molecular weight bioactive compound are: 1) the type of spacer and hydrolysable group between polymeric chain and immobilized bioactive ligand, 2) the level of common system polarity and its solubility in water, 3) the type of lyophilic groups, 4) the level of hydrophobic nature of the bioactive compound residue which is included in the phytoactive polymer, 5) the molecular mass of the polymer and 6) the configuration of the polymer s macromolecular structure in solution. For example, in the case of the polymeric derivatives of 1-amino-cyclopropane-1-carboxylic acid, when the ACC residue was attached to the polymeric carrier by various types of hydrolysable groups, it was found that the hydrolysis rate was strongly dependent on the type of hydrolysable group. In particular, increase of hydrolysis rate was

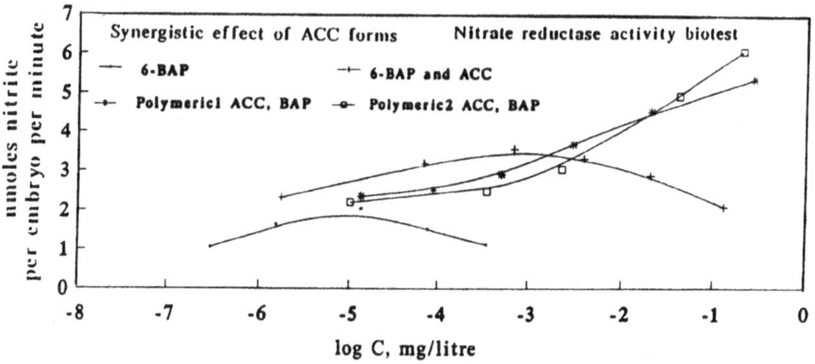

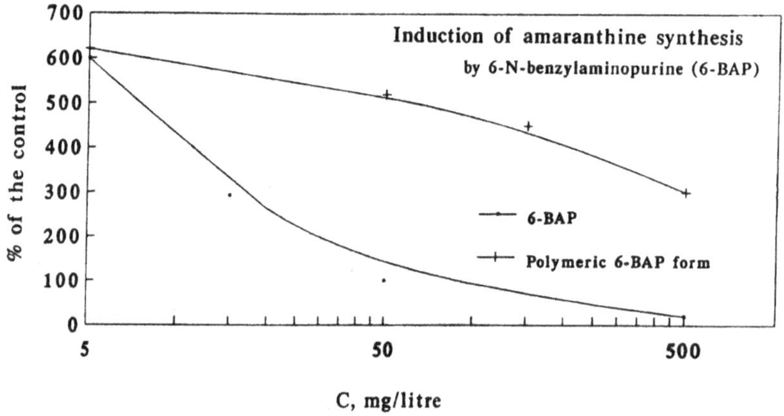

Figure 5. Upper: the expression differences of the synergistic effect of 1-aminocyclopropane-1-carboxylic acid and its polymeric forms (Kornakov *et al.*, 1992); lower: induction of amaranthine synthesis by 6-N-benzylaminopurine and its polymeric form.

linkages in the polymers, resulted in a notable shift in the stimulating concentrations to greater values. By introducing more hydrophilic substitutes in the polymer (as it was cited in the case of polymers of biologically active carbamates (Shtilman *et al.*, 1992), the region of stimulating concentrations shifted to a region of lower values.

A series of biotests was devoted to the influence of the phytoactive polymers on various *vital* function peculiarities of the plants. In particular, the affect of the phytoactive polymers with cytokininic activity on the reflective properties of the leaves and the water absorbance of the winter sort wheat (Shma'tko *et al.*, 1988), on the structure of the polar lipids of the chloroplast s membranes and their ultra-structure (Bocharova *et al.*, 1989; Grigoryuk *et al.*, 1990), and on the free-radical oxidation of lipids (Kolosha *et al.*, 1989) was studied. The influence of phytoactive polymer application on the content of endogenous cytokinins and adenosinephospates in the leaves of winter-sort wheat and potatoes and on other physiological processes of some cultures in conditions of water-stress (Grigoryuk *et al.*, 1990), on the content of chlorophyll in the leaves of wheat (Morozova *et al.*, 1986, 1989), on the mesostructure of potato leaves (Yakushkina *et al.*,1988), on the increase of maize stability in conditions of heightened salt earth (Guseinova *et al.*, 1991), on the germinating and growing power processes of seeds and sprouts of cereals (Allahverdiev *et al.*, 1991), on the damaging action of metals (ferrum, manganese, chromium) on sweet peaplant (Bessonova, 1991), on the stability of plants in conditions of low temperatures (Bocharova *et al.*, 1989; Trunova *et al.*, 1986) also investigated.

Phytoactive polymers with auxinic activity affected the accumulation of mineral substances in leaves of basma and henna plants in conditions of heightened chloride presence (Allahverdiev, 1991), to the functional protein content of henna leaves, the formation changes in root mitochondrions of rose sprouts (Grigoryuk *et al.*, 1991), to the germinating power of the tomato seeds in heightened salt conditions (Allahverdiev *et al.*, 1988), the regulation of the physiological processes in conditions of water-stress (Grigoryuk *et al.*, 1991), the growth of maize sprouts and plants (Ovoharov *et al.*, 1978), the increase of seed quality and crop yield of sugarbeet plants (Gontarenko, 1989), and the increase in productivity of essential producing crops and citrate crops (Ormotsadze *et al.*, 1991).

Data on biological activity of other types of phytoactive polymers are comparatively limited. The treatment of corn seeds with polyvinyl ester of nicotinic acid has resulted in notable growth stimulation (Ovcharov *et al.*, 1978). The application of the polymeric derivative of 2-chloroethylphosphonic acid to cereals has produced a remarkable retardant effect (Kurushina, 1991). The biologically activity of polymeric esters of gibberellic acid (Muromtsev *et al.*, 1990) and of the biological active carbamates (Shtilman, 1992) has been documented in standard biotests. The application of the polymeric form of fusicoccin on winter wheat has been shown to improve the wheat's frost-resistant properties (Muromtsev *et al.*, 1991). It has been demonstrated (Fig. 5) that polymeric forms of 1-aminocyclopropane-1-carboxylic acid produce a synergic stimulation on the nitrate reductase activity of amaranthus embryos, when applied together with cytokinins (e.g. BAP) higher than that of low molecular weight ACC (Kornakov *et al.*, 1992). The influence of the polymer structure to the biological activity (Figs 2 and 3), e.g. callus formation on leaf-segments of tobacco and grapevine genotypes (Paritsis *et al.*, 1992) has been also examined.

CONCLUSIONS

It may be concluded that even though investigations into the area of phytoactive polymers are relatively recent, a great number of representatives from this new generation group of biologically active high molecular weight compounds now have been synthesized. On the basis of the experimental data concerning the effect of the phytoactive polymers on plant life it may also be concluded that application of these polymers to agronomy is very promising.

It is clear that further investigations into new methods of polymeric PGR preparation, into the relation between the properties of polymeric PGRs and their structure and also into the biological activity of PGRs should be promoted.

ACKNOWLEDGEMENTS

We wish to thank Helene Tsatsakis-Pavlichenko for typing and preparing the manuscript and Mr Lukas Margiolakis for technical assistance.

REFERENCES

Akelah A, Selim A (1987) Application of functionalized polymers for controlled release formulations of agrochemicals and related biocides. *Chimica el Industria* 69:62-67.

Allahverdiev SR (1988) The influence of polystimulin on the growth and nitrate reductase activity of plants. *In*: Proc USSR Symp Plant Bio Polym, Nalchik, pp 68-69.

Allahverdiev SR, Abbasova ZI, Zeinalova EM (1988) The bioactivity of polystimulin A6 on basma and henna plants. *Ibid* pp. 21-22.

Allahverdiev SR, Zeinalova EM, Abbasova ZI (1991) Polystimulin A-6 formation changes in root mitochondrions of rose sprouts. *In*: Proc USSR Symp Plant Bio Polym, Zvinigorod, pp 66-67.

Allan GG (1967) Pesticide a liberation retades et son poroce de d'obtention. *Belg Patent* 706509.

Allan GG, Beer JW, Cousin MJ, Mikels RA (1980) The biodegradative controlled release of pesticides polymeric substrates. *In*: Controlled Release Technologies, Kydonieus AF (ed), CRC Press Inc, Boca-Raton, Florida.

Allan GG, Chopra CS, Neogi AN, Wilkins RM (1971) Design and synthesis of controlled-release pesticide-polymer combination. *Nature* 237:349-351.

Allan GG, Chopra CS, Neogi AN, Wilkins RM (1971) Controlled release pesticides. II. The synthesis of herbicide-forest solid waste combinations. *Tappi* 51:1293-1294.

Allan GG, Chopra GS, Friedhoff JF (1973) Pesticides pollution and polymers. *Chem Tech* 3:171-178.

Allan GG, Cousin MJ, Mikels RA (1979) Controlled release growth stimulants for plants. *Amer Chem Soc Polymer Prep* 20:311-343.

Allan GG, Stojanov A, Ueda M (1992) Controlled release stimulants *In*: Proc CRS Symp Bioactive Matrls, Kopecek J (ed), Orlando, CRS Press, 19:178-179.

Bessonova VP (1991) The influence of polystimulin K on plant photosynthetic apparatus in Fe, Mn and Cr medium. *Fiziol Biokhim Kulturn Rastenii (Kiev)* 23:158-164.

Bitner S, Perry I, Knobler Y (1979) Controlled release of plant hormones from natural and synthetic polymers. *Chimica el Industria* 61:291-297.

Bitner S, Perry I, Knobler Y (1977) Substained release of cytokinins from natural polymers. *Phytochemistry* 16:305-307.

Bocharova MA, Astahova NV, Trunova TI, Shtilman MI (1989) The increase of winter wheat frost-resistant properties by polystimulin K. *Fiziol Rastenii* 36:181-186.

Chailahyan MN (1982) Hormone regulation of plant growth. *Uspekhi Sovremennoi Biologii* 93:23-24.

Georgieva ES, Georgieva MP, Karanov E (1977) Possibility of prolonging the action of the growth regulating Allar Preparation. *Dokl Bolg Acad Sci* 30:1597-1600.

Georgieva MP, Georgieva E (1978) Uber die copolymerization von acrylamid mit vinylestern der 2,4-dichlorphenoxyessigsaure und 2,4-dichlorphenoxybuttersaure. *Angew Makromol Chem* 66:1-14.

Gontarenko SN (1989) The influence of polystimulin and chlorocholinechlorid on the seeds quality and crop yield of the sugarbeet plants. *In*: Priemi Povishenia Produktivnosti Fabrichnoi Saharnoi Svekli i Semennikov, Kiev, Inst Sugar Beet, pp 93-99.

Grigoryuk IA, Kurik MV, Shmat ko IG, Manuil'skii VD, Shtilman MI (1990) Influence of polystimulin K on the structure of polar lipids of winter wheat and potato chloroplast membranes with water deficite. *Fisiol Biokhim Kulturn Rastenii* 22:59-65.

Grigoryuk IA, Savinskii SB, Tkachev VI, Shmat'ko IG, Shtilman MI (1991) The influence of polystimulin K on winter wheat and potato. *In*: Proc USSR Symp Plant Bio Polym, Zvenigorod, pp 65-66.

Grigoryuk IA, Shmat ko IG, Kirichenko VP, Shtilman MI, Kornakov MYa (1990) The effect of polystimulin K on the yield and quality of winter wheat grain under bogharic and irrigated conditions. *Fisiol Biokhim Kulturn Rastenii* 22:5962.

Guseinova SG, Allahverdiev SR (1991) The influence of polystimulin K on the maize stability in conditions of heightened salt earth. *In*: Proc USSR Symp Plant Bio Polym, Zvenigorod, pp 62-63.

Harris FW, Dykes MR, Baker JA, Alabaugh AE (1977) Polymers containing pendent herbicide substituents Hydrolysis studies II. *ACS Symp Ser, Controlled Release Pestic. Symp* 53:102-111.

Harris FW, Post LK (1975) Synthesis of polymers containing pendent herbicide substituents. *Amer Chem Soc Polym Prep* 16:622-627.

Issa R, Ahelah A, Rehab A, Solaro R, Chiellini E (1990) Controlled release of herbicides bound to poly[oligo(oxyethylene)-methacrylate] hydrogels. *J Contr Rel* 13:1-10.

Jacubke HD, Bush E (1973) Synthese von kovalent an polymere trager gebundenen nie der molekularen wirkstoffen and versuche zur enzymatischen reaktivierung. *Zeitschrift fur Chem* 13:105-106.

Kellaway IW (1991) Controlled Release of Bioactive Materials. Proc 18th Int CRS Symp, Amsterdam, CRS press, Deerfield, IL.

Kidd H, Hartley D, Kennedy JM, James DR (1988) European Directory of Agrochemical Products, Plant Growth Regulators etc. Royal Soc Chem, Old Woking, Surrey.

Kogl F, Haagen-Smit AJ, Erxleben H (1934) The isolation of auxins *a* and *b* from plant materials. *Z Physiol Chem* 225:215-219.

Kolesnik YuR, Grib OK, Uskova LA, Svetkin YuV (1990) Plant growth activity of pesticides immobilized on poly-2-methyl-5-vinyl-pyridinemethylene-pentamethyl-disiloxan. *Fiziol Aktiv Vesh (Physiol Active Comp, Kiev)* 21:92-96.

Kolosha OI, Velikozhon LG, Ryaboklach VA (1989) The polystimulin K as regulator of the free radical oxidation of lipids and frost- resistance properties of plants. *Reports VASCHNIL* 5:5-7.

Kornakov MYa, Tsatsakis AM, Shtilman MI, Kurushina NV, Zalukaeva TP (1993) Polymeric forms of 1-aminocyclopropane-1-carboxylic acid. *J Contr Rel*, in press.

Korshak VV, Shtilman MI, Yaroshenko IV, Kravchenko YuA (1976) Synthesis and studies of the products of acylation of polyvinylalcohol by electric anhydride. *J High Molec Comp (VYSO SOED)* 18B:856-858.

Korshak VV, Shtilman MI, Yaroshenko IV, Bestuzheva TA (1977) Kinetics stydies of polyvinylalcohol acylation in solution. *Ibid* 19B:734-737.

Korshak VV, Shtilman MI, Mehtiev AH, Malisheva LI (1978) Polymeric esters of carboxylic acids. *Ibid* 20A:243-248.

Korshak VV, Shtilman MI, Mehtiev AH, Pomazanova AM (1979) Synthesis of polymeric esters on the basis of polyglicidylmethacrylate and some carboxylic acids. *Ibid* 21A:42-47.

Shtilman MI, Tsatsakis AM (1991) Phytoactive polymers Synthesis and peculiarities of their properties. *In*: *Macromol Chem, Macro Symp* (EPF Symp, Sorrento), suppl 48-49.

Tran Thanh Van K, Toubart P, Cousson A, Darvill AG, Gollin DJ, Shelf P, Albersheim P (1985) Manipulation of the oligosaccharides morphogenetic pathway by tobacco explants. *Nature* 314:615-617.

Trunova TI, Bocharova MA, Kuzina GV, Karnikova LD, Shtilman MI (1986) Polystimulin K activity on frost-resistant of winter wheat and fruit seeds. *Dokl AN SSSR* 286:508-512.

Tsatsakis AM (1986) Synthesis and investigations of aryloxy- arylmercapto carboxylic acids. PhD Thesis, Mendelejev Institute, Moscow.

Tsatsakis AM (1993) Controlled release pesticides. *Technika Chronika* TEE Publ, Athens, in press.

Tsatsakis AM, Mihalodimitrakis EN, Assithianakis P, Shtilman MI, Shashkova IB (1991) Phytoactive polymers Structure and properties dependence. *In*: Proc Int CRS Symp, Amsterdam, 18:550-551.

Tsatsakis AM, Roubelakis-Angelakis KA, Shtilman MI (1990) Synthesis and properties of plant growth regulating polymers. *In*: 2nd Congress in Chemistry of Greece and Cyprus, Athens A:180-184.

Tsatsakis AM, Shashkova IB, Shtilman MI (1992) Hydrolysis of polymeric esters of 1-naphthylacetic acid. *J High Molec Comp (VYSO SOED)* 34B:32-36.

Tsatsakis AM, Shtilman MI, Shashkova IM, Paritsis KN (1992) Hydrolysis of water soluble esters of 1-naphthyl acetic acid. *In*: Proc Int CRS Symp, Orlando, CRS Press, 19:443-444.

Tsatsakis AM, Dais Ph, Roubelakis-Angelakis KA (1993) Slow release auxins: Structure and biological activity relation. *In*: Panhellenic Congress in Chemistry, Athens, 14:78-82.

Wilkins RM (1982) The potential for the controlled delivery of plant growth regulators. *In*: Chemical Manipulation of Crop Growth and Development, McLaren JS, ed, Butterworth Sci, London, pp 111-126.

Yakushkina NI, Gurevich AS (1988) The significance of cytokinins in physiological action of Cu. *In*: Voprosi regulacii rostovih processov u rastenii, Moscow Govern Inst of Education, pp 3-8.

AUTHOR INDEX

Author	*Page*	*Author*	*Page*
Altamura, M	89	Marcé, M	113
Bagni, N	89	Marinho, P	221
Besford, RT	113	Marty, I	221
Biondi, S	89	Meeks-Wagner, D	161
Brugidou, C	221	Mengoli, M	89
Campos, JL	113	Meyer, Y	221
Capell, T	113	Osborne, DJ	1
Chartier, Y	221	Riera, R	113
Colon, A	55	Rivera, R	221
De Vries, SC	181	Roubelakis-Angelakis, KA	201
Engler, G	137	Segers, G	137
Engler, J	137	Shtilman, MI	259
Figueras, X	113	Tabler, M	237
Ferreira, P	137	Tiburcio, A	113
Fosket, D	55	Thorpe, TA	19
Gendy, CA	39	Tonoike, H	55
Grosset, J	221	Torrigiani, P	89
Han, IS	55	Tran Thanh Van, K	39
Hatjopoulos, P	145	Tsatsakis, AM	259
Hemerly, A	137	Van Engelen, FA	181
Inzé, D	137	Van Montagu, M	137
Kelly, AJ	161	Veylder, LD	137

SUBJECT INDEX

Abcisic acid, 25, 39, 72, 145, 259
Abcission, intertissue signalling, 12
  lacking mutants, 12
  zone, 9
Abs, 12
Acetabularia mediterranea, 3
Adiantum capullus-veneris, 104
Agrobacterium rhizogenes, 6, 99, 101
Agrobacterium tumefaciens, 25
Alnus
  glutinosa, 214
  incana, 214
Amyloplasts, 68, 222
Anantha, 8, 172
Angiosperm, 60
Antirrhinum majus, 175
Antisense RNA, 237
Apetala 1, 175
Apetala 2, 175
Apical, 3, 68, 101, 119, 142, 145, 162
Arabidopsis, 40, 59, 75, 137, 146, 226
Arabinogalactan proteins, 187
Arginine decarboxylase, 92, 115, 128
Aspergillus, 75
Assigning of gene function, 237
Auxin, 3, 22, 23, 39, 102, 142, 214, 222, 259
Avena sativa, 71, 114, 115, 128

Biological activity, 269
Brassica, 102, 130, 173, 209
Bulbocodium, 62
Brassimosteroids, 259

Caenorhabditis elegans, 75
Callose, 184
Callus, 19, 39, 169, 202, 221
Capillus veneris, 104
Capsicum, 115
Carrot, 4, 26, 40, 90, 139, 153, 191, 210
Carotenoid, 145, 247, 225
Catalase, 210
Catalytic RNA, 251
Chlamydomonas, 3, 6, 83
Cell
  clusters, 4, 23, 41, 89, 119
  cycle, 137, 162, 211
  cycle genes, 64, 139, 222
  density, 26
  division, 26, 41, 64, 89, 101, 137, 145, 161, 181, 204, 222
  differentiation, 118, 137, 161
  elongation, 63, 125, 161, 188, 205
  epidermal, 8, 40, 162, 210
  expansion, 5, 41, 70, 103, 182, 210
  layers, 23, 40, 169
  migration, 39, 137
  mobility, 145
  modification, 222
  polarity, 2, 67, 210
  population, 162, 221
  proliferation, 202, 227
  senescence, 161

controlled, 43
cytology of, 39
de novo, 22
gene expression, 147
in vitro, 22
inhibition of, 41
markers, 31, 39
molecular biology, 30
Ornithine decarboxylase, 92, 115
Oxidases, 186
Oxidative stress, 210

Passiflora, 96
Pathogenesis, 223
Pectins, 182
Pelvetia, 3
Pennisetum americanum, 204
purpureum, 206
Peptide, 9
Peripheral zone, 162
Peroxidases, 25, 44, 186, 210, 222, 225
Petiole, 11, 152
Petunia, 5, 24, 93, 185, 208
Picea glauca, 21
Pinus radiata, 21, 31
Pistil, 171
Pisum, 71
Phalaenopsis, 114, 119
Pharbitis, 164
Phaseolus vulgaris, 9, 123
Phenotypic mutant, 237
Photosynthesis, 223
Phragmoplast, 64
Phyla, 56
Physarum, 56
Physcomitrium eurystomum, 202
Phytoactive polymers, 259
Phytoalexins, 190
Phytohormones, 24, 259
Plasma membrane, 6, 207
Polarization, 4
Polyamines, 30, 39, 89, 113, 214, 259
putrescine, 47, 90, 115, 214
spermidine, 47, 50, 90, 115, 214
spermine, 47, 51, 92, 115, 214
Polymeric derivatives, 56, 263
PGR, 259
Polymers, 181, 259
Polysaccharide, 3, 41, 182, 208
Preprophase band, 64
Primordia, 4, 23, 40, 121, 145, 162
Procaryote, 213, 239
Phophocarpus tetragonolobus, 42
Protease, 225
Protein
binding, 39
extracellular, 180
glycine rich, 185
pathogenesis-related, 223
proline rich, 185
ribosomal, 225
stress, 43
synthesis, 9, 19, 56, 139, 146, 164, 222
Proteinaceous, 56
Protofilaments, 56
Protoplast, 114, 153, 201, 221
Protoplastization, 223
Prunus avium, 96
Pseudogenes, 58
Psophocarpus terragonolobus, 42
Pulvinus, 9
Putrescine, 47, 90, 97, 115, 214

Radial array, 64
Recalcitrance, 117, 202
Rhamnogalacturonan, 182
Rhizoid, 3
Rhizogenesis, 7, 96, 202
Ri-T-DNA, 102
Ribosome, 126
Ribozymes, 237
RNA synthesis, 29, 104, 126
Root, 7, 19, 45, 59, 100, 121, 142, 145, 170, 221

Saccharomyces cerevisiae, 137, 139

Sampucus nigra, 9
Schizosaccharomyces pombe, 137
Secondary zones, 10
Serine protease inhibitor, 1
Shoot apex genes, 142
Shoot meristem, 161
Silene coeli-rosa, 169
Sinapis alba, 5, 114, 145
Spermidine, 47, 50, 90, 115, 214
Spermine, 47, 92, 115, 214
Starch, 28, 30
Squamosa, 175
Sorghum, 41
Soybean, 58, 100, 139, 185, 227
Stratiotes aloides, 201
Superoxide dismutase, 210

Taxus, 62
Thin Cell Layer, 43
Thionins, 187
Thioredoxins, 227
Thylakoid, 123, 221
Thymidine, 222
Tomato, 8, 172
Tradescantia, 61
Trans-acting, 153, 239
Transcriptional, 125, 146
Trans-factors, 153
Transgenic, 99, 141
Transglycosylases, 186
Trigonella, 115
Triticum monococcum, 116
Tubulin, 57
  genes, 74
  isotypes, 75

Ubiquitin, 57, 224

Vacuheria, 201
Variable phenotypes, 241
Vegetative, 1, 90, 114, 148, 161
Vertebrates, 61
Vicia faba, 6
Vigna, 114, 115
  aconitifolia, 118
Viroid, 251
Virus, 250
Vitis vinifera, 92, 205

Xanthium
  pennsylvanicum, 162
  strumarium, 105
Xenopus, 125, 141, 239
Xylans, 184
Xyloglucan, 184

Zea mays, 114, 119, 139

GPSR Compliance

The European Union's (EU) General Product Safety Regulation (GPSR) is a set of rules that requires consumer products to be safe and our obligations to ensure this.

If you have any concerns about our products, you can contact us on ProductSafety@springernature.com

In case Publisher is established outside the EU, the EU authorized representative is:

Springer Nature Customer Service Center GmbH
Europaplatz 3
69115 Heidelberg, Germany

Batch number: 10761599

Printed by Printforce, the Netherlands